PROTEIN BIOCHEMISTRY, SYNTHESIS, STRUCTURE AND CELLULAR FUNCTIONS

PROTEOMICS: METHODS, APPLICATIONS AND LIMITATIONS

PROTEIN BIOCHEMISTRY, SYNTHESIS, STRUCTURE AND CELLULAR FUNCTIONS

Additional books in this series can be found on Nova's website under the Series tab.

Additional E-books in this series can be found on Nova's website under the E-books tab.

PROTEIN BIOCHEMISTRY, SYNTHESIS, STRUCTURE
AND CELLULAR FUNCTIONS

PROTEOMICS: METHODS, APPLICATIONS AND LIMITATIONS

GISELLE C. RANCOURT
EDITOR

Nova Science Publishers, Inc.
New York

For permission to use material from this book please contact us:
Telephone 631-231-7269; Fax 631-231-8175
Web Site: http://www.novapublishers.com

Additional color graphics may be available in the e-book version of this book.

LIBRARY OF CONGRESS CATALOGING-IN-PUBLICATION DATA

Proteomics : methods, applications and limitations / editor, Giselle C.
Rancourt.
p. ; cm.
Includes bibliographical references and index.
ISBN 978-1-61668-691-8 (hardcover)
1. Proteomics. I. Rancourt, Giselle C.
[DNLM: 1. Proteomics--methods. 2. Biological Markers. 3. Celiac
Disease--diagnosis. 4. Models, Genetic. 5. Proteome--diagnostic use. 6.
Proteome--genetics. QU 58.5 P96747 2010]
QP551.P75672 2010
572'.6--dc22 2010007621

Published by Nova Science Publishers, Inc. † New York

CONTENTS

PREFACE

Proteomics is the large-scale study of proteins, particularly their structures and functions. Proteins are vital parts of living organisms, as they are the main components of the physiological metabolic pathways of cells. The proteome is the entire complement of proteins, including the modifications made to a particular set of proteins, produced by an organism or system. This book presents and discusses various topical data on proteomics, including: proteomic technologies used in celiac disease, protein location in plant proteomics, the prediction of protein-protein interactions (PPIs) using proteomics, proteomic studies of mild cognitive impairment, among others.

Chapter 1 - Proteomic technologies are used with increasing frequency in the scientific community. In this review the authors would like to highlight their use in celiac disease. The available techniques that include two-dimensional gel electrophoresis, mass spectrometry, antibody and tissue arrays, have been used to identify proteins or protein expression changes specific of gut tissue from patients with celiac disease. A number of studies have employed proteomic methodologies to look for diagnostic biomarkers in body fluids or to examine protein expression changes and posttranslational modifications during signaling. The fast technological development of technologies, along with the combination of classic techniques with proteomics, will lead to new discoveries which will consent a better understanding of the pathogenesis of celiac disease and its complications (i.e. refractory CD and cancer), and to possibly indicate targets for an early diagnosis of CD complications and for specific terapeutic approaches.

Chapter 2 - Organelle proteomics allows the characterization of complex proteomes to understand the protein networks which regulate growth and development, as well as adaptation and evolution. Purification of organelles is of paramount importance and diverse protocols are published. Some organelles such as chloroplasts, mitochondria, and the nucleus are surrounded by membranes which facilitate their purification. Others have membranes easily disrupted (vacuoles and peroxisomes), or are complex systems for protein trafficking (endoplasmic reticulum, Golgi, and secretory vesicles). The cell walls present different difficulties since they have no physical limits allowing purification. The purity of the targeted cell compartment is usually evaluated by biochemical and/or immunological methods. Nevertheless, in any sub-cellular proteomic analysis, proteins from a different compartment can be detected and the difficulty is to decide whether it is a contamination, or the unexpected location is real and has a functional significance. Software to predict sub-cellular location of proteins is available. However, since not all the targeting signals are known at present, carefulness in the use of such tools is recommended. Different tactics to solve this puzzle are discussed in this commentary.

Chapter 3 - The simultaneous analysis of all proteins expressed by a cell, tissue or organism in a specific physiological condition is the main goal of proteomic studies. Gel-based proteomic is the most popular and versatile method of global protein separation and quantification. This is a mature approach to screen the protein expression at the large scale. Based on two independent biochemical characteristics of proteins, two-dimensional electrophoresis combines isoelectric focusing, which separates proteins according to their isoelectric point, and SDS-PAGE, which separates them further according to their molecular mass. The next typical steps of the flow of gel-based proteomics are spots visualization and evaluation, expression analysis and finally protein identification by mass spectrometry. At present, two-dimensional electrophoresis allows simultaneously to detect and quantify up to thousand protein spots in the same gel in a wide range of biological systems for the study of differentially expressed proteins. However, gel-based proteomic has a number of inherent drawbacks. In this review article, the benefits, difficulties, limits and perspectives of gel-based proteomic approaches are discussed.

Chapter 4 - In this commentary, the authors review algorithms for the analysis of liquid chromatography-mass spectrometry (LC-MS) data. Mass spectrometry is a technology that can be used to determine the identities and abundances of the compounds in complex samples. In combination with liquid chromatography, it has become a popular method in the field of proteomics, the large-scale study of proteins and peptides in living systems.

The data sets obtained from an LC-MS experiment are large and highly complex. The outcome of such an experiment is called an LC-MS map and is a collection of mass spectra. They contain, among the signals of interest, a high amount of noise and other disturbances. That is why algorithms for the low-level processing of LC-MS data are becoming increasingly important.

In this commentary, the authors revied the state-of-the-art of quantification algorithms, their capabilities and also limitations and outline avenues for future research.

Chapter 5 - Nowadays, one of the most important goals of Proteomics is the prediction of protein-protein interactions (PPIs), whose knowledge is vital for all biological processes. In the present paper the authors propose an approach to the prediction of protein-protein interactions in yeast based on the well-known paradigm of Support Vector Machines (SVM) for classification and feature selection methods using Genomics/Proteomics information from the main databases. In order to obtain higher values of specificity and sensitivity for their predictions, the authors took a high reliable set of positive and negative examples for the construction of the SVM model. The authors then extracted a set of proteomic/genomic features from these examples and also introduced a similarity measure in the calculation of the features that allow us to improve the prediction capability of their model. In the analysis of the results, the authors also applied their approach to in vitro datasets, obtaining high accuracy classifications. Their final SVM classifiers obtain a low error rate in the prediction for each pair of proteins of several datasets for both in vitro and in silico methodologies.

Chapter 6 - Recent innovations in liquid chromatography-mass spectrometry (LC-MS) based methods have facilitated comparative and functional proteomic analyses of large numbers of proteins derived from complex samples without any need for protein or peptide labelling. Here the authors discuss the features of label-free LC-based proteomics techniques. The authors first summarize recent methods used for quantitative protein analyses by MS techniques. The major challenges faced by label-free LC-MS based approaches are discussed; these include sample preparation, peptide separation, data mining and quantification. Absolute quantification, kinetic approaches and database search algorithms are also addressed. The authors focus on the Expression^{E} System^{TM} (Waters, Manchester, UK), a relatively new platform allowing label-free quantification of peptides for which mass and retention time have been accurately measured. Enhancing the power of this method will require developments in both separation technology and bioinformatics/statistical analysis.

Chapter 7 - Mild cognitive impairment (MCI) is arguably the earliest form of Alzheimer's disease (AD). Better understanding of brain changes in MCI may lead to the identification of therapeutic targets to slow the progression of AD. Oxidative stress has been implicated as a mechanism associated with the pathogenesis of both MCI and AD. In particular, among other markers, there is evidence for an increase in the levels of protein oxidation and lipid peroxidation in the brains of subjects with MCI. Several proteins are oxidatively modified in MCI brain, and as a result individual protein dysfunction may be directly linked to these modifications (e.g., carbonylation, nitration, modification by HNE) and may be involved in MCI pathogenesis. Additionally, Concanavalin-A-mediated separation of brain proteins has recently led to the identification of key proteins in MCI and AD using proteomics methods. This chapter will summarize important findings from proteomics studies of MCI, which have provided insights into this cognitive disorder and have led to further understanding of potential mechanisms involved in the progression of AD.

Chapter 8 - The general strategy in proteomic research consists in sample preparation, protein or peptide separation, their identification, and data interpretation. A critical step is certainly protein or peptide separation. Since increasingly complicated biological structures are studied by mass spectrometry (MS), the need for more powerful and highly resolving separation methods is growing. Consequently, multidimensional separation techniques in combination with MS have emerged as a powerful tool for the large-scale proteomic analysis. Until recently, two dimensional gel electrophoresis (2-DE) was the technique most often used for protein separation. The limitations of 2-DE in detecting low abundance, very small or large proteins, basic and membrane/hydrophobic ones, as well as difficulties with process automation, have forced researchers to look for other methods of protein separation, such as multidimensional liquid chromatography coupled to MS (MDLC-MS) or tandem MS (MDLC-MS/MS). MDLC combines two or more forms of LC to increase the peak capacity, and thus the resolving power of separation, to better fractionate peptides prior to entering the mass spectrometer. In this chapter, the authors analyze status and recent developments of the MDLC experiments in their fundamental components. It describes a variety of separation modes that have been employed to achieve protein-level or peptide-level separation, including size exclusion chromatography, ion exchange chromatography, and reversed-phase chromatography. The authors also discuss the advantages and disadvantages of two different approaches that can be followed for the studies of proteomics: protein-level separation or peptide-level separation.

Commentary - It is well established that high doses of ionising radiation, such as used in radiotherapy, increase risk of cardiovascular diseases (CVD). Observed effects include direct damage to the coronary arteries, marked diffuse fibrotic damage of the pericardium and myocardium, pericardial adhesions, stenosis of the valves and microvascular damage. In contrast, there are considerable uncertainties concerning health effects of low doses of ionising radiation on heart. The need to explore potential biological and physiological effects at low doses is being increasingly acknowledged as the plans for new nuclear power plants and novel medical applications using low-dose radiation are emerging.

In: Proteomics
Editor: Giselle C. Rancourt
ISBN: 978-1-61668-691-8

Chapter 1

PROTEOMICS IN CELIAC DISEASE

V. De Re, MP. Simula, L. Caggiari, A. Pavan , V. Canzonieri and R. Cannizzaro

Farmacologia Sperimentale e Clinica DOMERT, Centro di Riferimento Oncologico, IRCCS, Aviano, Pordenone, Italy

ABSTRACT

Proteomic technologies are used with increasing frequency in the scientific community. In this review we would like to highlight their use in celiac disease. The available techniques that include two-dimensional gel electrophoresis, mass spectrometry, antibody and tissue arrays, have been used to identify proteins or protein expression changes specific of gut tissue from patients with celiac disease. A number of studies have employed proteomic methodologies to look for diagnostic biomarkers in body fluids or to examine protein expression changes and posttranslational modifications during signaling. The fast technological development of technologies, along with the combination of classic techniques with proteomics, will lead to new discoveries which will consent a better understanding of the pathogenesis of celiac disease and its complications (i.e. refractory CD and cancer), and to possibily indicate targets for an early diagnosis of CD complications and for specific terapeutic approaches.

Keywords: two-dimensional gel electrophoresis: DIGE; mass spectrometry, MALDI-TOF; T-cell lymphoproliferations, lymphomas.

Proteomic Approaches

Proteomics uses a rapidly evolving group of technologies to identify, quantify, and characterize a global set of proteins. In this review, we will highlight important advances in understanding Celiac Disease (CD) using proteomic technologies and suggest future directions. It is not their intention to present an exhaustive review of proteomics in CD but rather to highlight the specific studies as examples of a possible methods to better understand the pathogenesis of celiac disease and its complications and possibly to indicate prognostic markers and targets for specific terapeutic approaches.

The word proteomics was coined in 1994. Initial studies demonstrated that a large number of gut and serum proteins could be visualized on a two dimensional electrophoresis (2DE) gel, however only with the progress of mass spectrometry (MS) and informatics, in the early 1990s, that protein identification from gels became routine. More recently, a number of other techniques using MS, chromatography, and protein affinity surfaces have evolved, but still the use of proteomic tools in CD research remains limited.

Proteomic techniques can be grouped according to two different approaches dipending on whether intact proteins or proteins digested into peptides are analysed. 2DE is the most widely used methods in which the intact proteins are separated before digestion and identification. This technique separates proteins according to isoelectric point and molecular weight, and proteins are visualized and quantified by staining. More recently, the 2DE has been improved by labelling two samples and a pooled internal standard with different fluorescent dyes. In this technique, called fluorescence difference in gel electrophoresis (DIGE), two samples, and the internal standard, are simultaneously visualized in the same gel because of their discrete excitation and emission wavelengths, thus improving both the reproducibility and the ability to quantify proteins. 2DE, coupled with peptide mass fingerprinting and the subsequent database analysis of spectra data, allows protein identification. DIGE represents a major advance in this 30- year-old technique and it is now widely employed.

Other proteomic techniques first digest the sample and then separate the peptides for identification: like in the liquid chromatography/MS (LC/MS), in which digested peptides are separated first by liquid chromatography and then injected directly into the mass spectrometer. LC/MS can identify low-abundance and hydrophobic proteins that are not detected by 2DE. Moreover, with the use of isotopic tags, LC/MS can also quantify proteins. Isotope-coded affinity tags (ICAT) contain three functional regions: an affinity purification region, a peptide-binding one and an isotopically distinct linker region. A biotin tag is used for affinity purification. A thiol-specific binding moiety is used to covalently link the reagent to cysteine residues in a target peptide. The intervening linker region is isotopically labeled with either $_{12}C$ or $_{13}C$, so that peptides labeled with the reagent are chemically identical but can be distinguished in a mass spectrometer based on their mass differences. This advance allowed the mass spectrometer to be used to quantify protein abundance differences in two samples but it has several disadvantages including labeling only peptides containing amino acid cysteine. More recently, a similar technique called iTRAQ, which labels all peptides, was introduced: it allows increased confidence in the identification of proteins, because multiple peptides for the protein are identified, and it also permits the simultaneous quantification of samples. This technique has been applied to biomarker discovery in plasma. It consists in a non-gel based technique with the purpose to identify and quantify proteins from different sources in one single experiment. For example, the method is based on the covalent labeling of the N-terminus amines of peptides from protein digestions with tags of varying mass. When up to four differently labeled peptide mixtures are combined, it allows the identification of the source mixture of every peptide. The combined labeled are usually fractionated by nano LC and analyzed by tandem MS (MS/MS). A database search is then performed by using the fragmentation data to identify the labelled peptides and hence the corresponding proteins. The fragmentation of the attached tag generates a low molecular mass reporter ion that can be used to relatively quantify the peptides and the proteins from which they originated.

Newer approaches have also allowed for identification of posttranslational modifications by MS as well. Surface-enhanced laser desorption/ionization (SELDI) MS measures protein ions after the proteins are selectively bound to a plate coated with an affinity surface. SELDI MS measures the intensity of peaks from the subset of proteins captured on the plate. Peak height differences between samples correlate with a relative abundance in the sample. The identification of the protein represented by the peak can be difficult.

Capillary electrophoresis coupled to MS (CE/MS) provides high separation efficiency in small volumes. Proteins are resolved according to their elution time from the CE and the mass of the protein in the MS.

Protein-binding arrays use antibodies or synthetic protein-binding small molecules, like peptoids or aptamers, to visualize protein binding. Antibody arrays are not an unbiased discovery analysis since they only measure a fixed set of proteins but can provide multiplexed measurements of protein abundance.

Each technique presents advantages and disadvantages: all have been used for the analysis of CD to identify the presence of proteins, to compare their abundance between different CD-samples, and to analyze changes in posttranslational modifications. The improvements in quantification and identification of modified proteins will facilitate better understanding of gut physiology.

A brief description of the matters in which these techniques mostly find their application and a mention of the principal results obtained by proteomics approaches in CD are reported below.

Identification and Characterization of Gliadin-Derived Epitopes by Mass Spectrometry

Celiac disease is a common severe intestinal disease resulting from intolerance to dietary wheat gluten and related proteins. The large majority of patients present the HLA-DQ2 and or DQ8 molecules. Gluten-specific HLADQ-restricted T cells have been found at the site of the lesion only in the gut of CD patients. The nature of peptides that are recognized by such T cells, however, has been so far unclear. The principal toxic components of wheat gluten belong to a family of closely related proline and glutamine rich proteins called gliadins. The characterization of gliadin-derived peptides that are primarily recognized by intestinal gluten-specific HLA-DQ-restricted T cells is a key step towards the development of strategies to interfere in mechanisms involved in the pathogenesis of celiac disease. An enzymatic digestion (for ex. rat jejunum brush border enzymes) has been performed to obtain gliadin hydrolysis products and their subsequent analysis by LC-MS [1]. The LC-separated peptides were further fragmented and detected by electrospray ionization mass spectrometry. The compounds defined by their m/z values were detected by their ionization intensities. The amino acid sequences of the proteolytic products were determined from the MS-MS fragmentation pattern. Then T-cell proliferation assays that use the identified gliadin peptides coupled with antigen presenting cells and specific T-cell clones established from CD intestinal biopsy samples were performed to identify major peptides involved in CD. By using mass approaches, to date at least 17 distinct DQ2 restricted T cell epitopes have been identified from gluten proteins found in disease associated grains, reviewed in [2]. The 33-mer peptide LQLQPFPQP-

QLPYPQPQLPYPQPQLPYPQPQPF and the 25-mer peptide FLQPQQPFP-QQPQQPYPQQPQQPFPQ are recognized as the major celiac-toxic segments present in α-2 gliadin and γ-gliadin. Both these peptides are found resistant to gastric, pancreatic and intestinal brush border degradation and were a good substrate of human transglutaminase 2 (TG2) and of DQ2 T –cell activation. The complex pattern of the sample digests, evidenced by MS, was indicative of a broadly heterogeneous mixture harbouring, in addition to the 25- and 33-mer, a still undetermined number of epitopes deserving further structural investigation with the same analytical approach.

Quantification of Gliadin and Gliadin-Derived Epitopes by Mass Spectrometry

Presently, the only treatment for CD consists in a life-long gluten-free diet (GFD). Several attractive targets for new CD treatments are under investigation. Complementary strategies aiming at interfering with the activation of gluten-reactive CD4-T cells include the inhibition of intestinal TG2 activity to prevent the selective deamidation of gluten immunogenic peptides [3] and the blockage of the binding of gluten epitopes to the HLA-DQ2 and HLA-DQ8 molecules [4]. Other treatments include cytokine therapy [5] selective adhesion molecule inhibitors [6], and peptide degradation by prolyl endopeptidases (PEPs) of microbial origin [7-8]. Proteomic technologies were used to verify the exclusion of toxic epitopes during food processing. The pretreatment of gluten with lactobacilli and fungal proteases are examples [9]. Pretreated-food prevented the development of fat or carbohydrate malabsorption in the majority of CD patients. Proteomic approach consents to identify new organisms that efficiently degrade gluten T-cell-stimulatory peptides by cytoplasm internalization and degradation. Moreover, mass spectrometry-based approaches could be used to quantify gliadin and/or its degradated peptides in the food and after food pretreatment, respectively [10].

Research of Substracts for Transglutaminase

TG2 is a calcium-dependent enzyme that catalyzes the formation of covalent bonds between free amine groups in one protein and protein-bound glutamines of another, creating highly cross-linked protein complexes. TG2 is ubiquitously expressed and has multiple normal physiologic functions, such as blood clotting, wound healing, cell adhesion, apoptosis and barrier formation. TG2 has also been associated with certain pathologic conditions [i.e., inflammatory diseases, such as encephalomyelitis, inflammatory myopathies, and celiac disease, as well as various types of cancer] [11].

Apart from the deamidation of gliadin peptide, TG2 acts as hapten in the generation of antibodies against gliadin and itself, catalyzing the synthesis of heteromeric gliadin-TG2 complexes that may provoke an immune response to TG2 by stimulating normally silent autospecific B-cells [12]. Moreover, it is known that CD is frequently associated with several other autoimmune disorders, such as autoimmune thyroid disease, type 1 diabetes mellitus, autoimmune liver diseases and inflammatory bowel disease [13]. It is supposed that the exposure of self-antigens deriving from TG2 interaction may lead to the development of autoantibodies [14]. For what regards specifically the CD, it remains elusive how TG2, mainly a cytoplasmatic protein, comes into contact with gliadin, which is in the intestinal lumen. Noably, it is now established that CD patients exhibit an increased intestinal permeability that allows gliadin to have access to the subepithelial compartment, and that this is an early feature of the disease process [15]. It has also been hypothesized that, either TG2 is constitutively in the open catalytically active conformation in the celiac mucosa or that more likely gluten triggers extracellular TG2 activation by interacting with certain innate immune receptors, as TLR-2, −3, and −4, that were found up-regulated in the celiac mucosa, even in patients whose disease is in remission [16]. The occurrence of an innate immune response to gluten in CD patients involving NKG2D/MICA receptor and IL-15 cytokine have also been reported [17, 18]. More recently, it has been demonstrated that α_2-gliadin-33mer may translocate by transcytosis through the gut epithelium and that this process is regulated up to 40% by interferon-γ administration [19].

A proteomic approach has been used to identify TG-modified protein targets in human intestinal epithelial cells. In short, experimental procedure used biotinylated protein substrates as probes, which are covalently incorporated into cellular proteins by TG2 in a calcium-dependent reaction [20]. Cells were lysed, and biotinylated proteins extracted by affinity chromatography using streptavidin-alkaline phosphatase; the excess probe was eliminated by gel filtration. The biotin-labeled protein mixture was separated by Sodium Dodecyl Sulphate-PolyAcrylamide Gel Electrophoresis (SDS-PAGE. Proteins were then digested in gel by trypsin and identified using the

MALDI-time-of-flight (MALDI-TOF) mass spectrometer and Data Base Searching. A collection of the transglutaminase substrate proteins and interaction partners is accessible in the TRANSDAB database (http://genomics.dote.hu/wiki/index.php/).

With respect to their biological significance, the identified proteins fall into four groups. These targets include proteins involved in cytoskeletal network organization, in the folding mechanism, in the transport process and in miscellaneous metabolic functions such as phosphoglycerate dehydrogenase, a key enzyme in lipid metabolism.

Cytoskeletal proteins cross-linked with TG2 proteins are essentially found to be involved in the apoptosis cell death [21]. In brief, the findings support the hypothesis that the post-translational modifications of proteins involved in cytoskeletal rearrangement are needed for maintaining tissue integrity and to influence the interactions with other proteins important for enterocyte survival [22]. Among all these proteins, actin, by generating anti-actin antibodies, is shown to be strongly associated with more severe degrees of intestinal damage (3a, 3b, 3c) [23].

The second group of proteins represents proteins with chaperone activities (Hsps). The expression of constitutively Hsps, such as Hsp60, Hsp70, Hsp72, and Hsp90 and Bip, by enterocytes may be part of a protective mechanism developed by the intestinal epithelium to treat harmful components present in the intestinal lumen. Three genes of the Hsp 70 family are located in the MHC class III region, and are particularly intersting since the Hsps seem to be involved in the antigen processing and presentation pathway and are thought to play a role in the pathogenesis of some autoimmune systemic disorders [24]. Recently we found the same AH8.1 ancestral haplotype variant in two CD subjects, among 43 patients tested, from the same geographical area; the same haplotype was absent in 70 blood donors tested [25]. One of the patients with AH8.1 variant presented this haplotype in homozogosis and was affected by several autoimmune diseases (thyroiditis, myalgia, diabete mellitus and CD), suggesting a role of HLA class III in the autoimmune predisposition.

The last groups consisted of proteins involved in several transport processes and miscellaneous metabolic pathways. For example it is reported that TG2 interacts with proteins implicated in the correct assembly of the respiratory chain complexes into the mitochondria, as well as with proteins inducing the release of apoptosis.

It is intriguing to suggest that TG2 could come in contact with gliadin inside enterocytes and can modify gliadin peptides by crosslinking to itself or to other acyl-acceptor substrates thus originating neo-antigens recognized by the immune system. This mechanism could explain the existence of auto-antibodies in CD with several distinct specificities.

Human Body Fluid Proteome Analysis

The ability to determine global identification and quantification of body fluid proteins, may provide useful clinical diagnostic and prognostic information and the elaboration of putative biomarkers for a variety of human diseases. In terms of disease diagnosis and prognosis, a human body fluid (*e.g.*, blood, urine, or saliva) appears to be more useful because body fluid testing provides several key advantages including low invasiveness, minimum cost, and easy sample collection and processing. Moroever, autoantibodies appear to be very useful for analysis and characterization of serum autoantigens in autoimmune diseases.

The presence of immunoglobulin A (IgA) antiendomysial antibodies is specific for celiac disease and is used for screening, diagnosis and follow-up of this disease with a high sensitivity and specificity [26]. The major antigen target of IgA antiendomysial antibodies was identified as TG2 [27]; nevertheless, proteome analysis has been performed to search for additional celiac disease-specific autoantigens. To this aim CD cell proteins were separed on two-dimensional gel electrophoresis gels, then a western blot with patients and healthy subjects sera, and identification of CD-specific detected antigens was performed by MS. Alternatively multiple affinity protein profiling combines antibodies from CD patients, immobilized to a resin on the Fc region, to enrich proteins of interest from sera followed by elution of captured proteins and MS identification. Peptides were analyzed by comparison from protein profile obtained with antibodies from blood donors.

Using these proteomic approaches some proteins that were immunorecognized with various frequencies by sera of CD patients have been reported. Four are autoantigens: actin and ATP synthase βchain and two variants of enolase α with possible diagnostic utility [28]. Clearly, a validation of these autoantibodies will be carry out, nonetheless, the result reported is suggestive of the existence of additional CD autoantigens along to TG2.

Differential in Gel Protein Expression in CD Patients and Controls

The overall gut epithelium protein expression in CD was poorly investigated; new findings are now available by confronting the average of the protein expression from biopsy-proven untreated CD patients and controls using the 2D-DIGE approach [29].

Altogether, data indicate a down regulation of proteins belong to the PPAR signalling pathway, as HMGCS2, FABP1, FABP2, PCK2, APOC3, ACADM. Targets of PPAR signaling regard primarily fatty acid oxidation and lipid metabolism, and secondary inflammation and induction of apoptosis, adipogenesis, and glucose control [30]. Much attention has focused on the role of the different PPARs in the human intestine, in particular on their importance in inflammation [31] and neoplastic transformation [32-34]. Fatty acids, and particularly their eicosanoid products, resulting from the lipoxygenase, cyclooxygenase, and P450 pathways, show high affinity for PPARs, and some of them have been suggested as endogenous PPARα ligands. Recently an association of the CYP4F2 and CYP4F3 genes with inflammatory celiac disease have been reported [35]. These genes are included in the cytochrome P450 gene 4 family (CYP4) consisting of a group of over 63 members. Several human diseases have been genetically linked to the expression CYP4 gene polymorphic variants, which may link human susceptibility to diseases of lipid metabolism and the activation and resolution phases of inflammation. In particular, CYP4F2 and CYP4F3 catalyse the inactivation of leukotriene B4 (LTB4), a potent mediator of inflammation responsible for recruitment and activation of neutrophils. The association of LTB4-regulation with innate immune response of neutrophil mobilization is now connected with the established Th1 adaptive immunity present in coeliac disease patients [36].

Secondarily but still important, from our proteomic research, the gut IgM expression appears the best marker associated to a clear CD condition. The IgM secretion has previously reported associated to CD [37] and since IgM antibodies can activate complement, it is suggested that they might contribute to the damage following an encounter with antigens (e.g. gluten) [38, 39]. It is known that IgA and IgG classes in serum (systemic immunity) and of IgA and IgM classes in jejunal aspirate and gut lavage fluid (mucosal immunity) response occur in untreated CD. Enhanced local secretion of IgA ($p<0.05$) and IgM ($p<0.001$) with respect to controls has also been demonstrated in CD patients using *in vitro* lymphocyte culture [37]. Our data confirm the concept of an IgM segregation in the gut, and indicate that IgMs and B cells are important for CD immunopathogenesis.

Conclusion

Proteomic approaches are only beginning to be applied to the study of celiac disease. In this study we reported some insight into the strengths of emerging proteomic applications in the CD studies. Major efforts have been found in the research of gliadin-epitopes, for the substracts of transglutaminase, for the identification of diagnostic markers, and for the pathways associated to the CD disease. In these fields proteomics approaches provide a promising tool not routinely used in the laboratory. In the near future these powerful approaches might be used as a standard technique for diagnosis of CD, and identification of biomarkers useful for target therapies.

References

[1] Hausch, F; Shan, L; Santiago, NA; Gray, GM; Khosla, C. Intestinal digestive resistance of immunodominant gliadin peptides. *Am J Physiol Gastrointest Liver Physiol*, 2002 283, G996-G1003.

[2] Ferranti, P; Mamone, G; Picariello, G; Addeo, F. Mass spectrometry analysis of gliadins in celiac disease. *J Mass Spectrom*, 2007 42, 1531-1548.

[3] Molberg, O; McAdam, S; Lundin, KE; Kristiansen, C; Arentz-Hansen, H; Kett, K; Sollid, LM. T cells from celiac disease lesions recognize gliadin epitopes deamidated in situ by endogenous tissue transglutaminase. *Eur J Immunol*, 2001 31, 1317-1323.

[4] Kim, CY; Quarsten, H; Bergseng, E; Khosla, C; Sollid, LM. Structural basis for HLA-DQ2-mediated presentation of gluten epitopes in celiac disease. *Proc. Natl. Acad. Sci.* 2004 101, 4175-4179.

[5] Salvati, VM; Mazzarella, G; Gianfrani, C; Levings, MK; Stefanile, R; De Giulio, B. Recombinant human IL-10 suppresses gliadin dependent T cell activation in ex vivo cultured celiac intestinal mucosa. *Gut* 2005 54, 46-53.

[6] Sollid, LM; & Khosla, C. Future therapeutic options for celiac disease. *Nat. Clin. Pract. Gastroenterol. Hepatol.* 2005 2, 140-147.

[7] Gass, J; Bethune, MT; Siegel, M; Spencer, A; Khosla, C. Combination enzyme therapy for gastric digestion of dietary gluten in patients with celiac sprue. *Gastroenterology*. 2007 133, 472-480.

[8] Mitea, C, Havenaar, R; Drijfhout, JW; Edens, L; Dekking, L; Koning, F. Efficient degradation of gluten by a prolyl endoprotease in a gastrointestinal model, implications for coeliac disease. *Gut*. 2008 57, 25-32.

[9] Rizzello, CG; De Angelis, M; Di Cagno, R; Camarca, A, Silano, M; Losito, I; De Vincenti, M; De Bari, MD; Palmisano, F; Maurano, F; Gianfrani, C; Gobbetti, M. Highly efficient gluten degradation by lactobacilli and fungal proteases during food processing, new perspectives for celiac disease. *Appl Environ Microbiol*. 2007 73, 4499-4507.

[10] Marti, T; Molberg, O; Li, Q; Gray, GM; Khosla, C; Sollid, LM. Prolyl endopeptidase-mediated destruction of T cell epitopes in whole gluten, chemical and immunological characterization. *J Pharmacol Exp Ther*. 2005 312, 19-26.

[11] Caputo, I; D'Amato, A; Troncone, R; Auricchio, S; Esposito C. Transglutaminase 2 in celiac disease. *Amino Acids*. 2004 26, 381-386.

[12] Stenman, SM; Lindfors, K, Korponay-Szabo, IR; Lohi, O; Saavalainen, P; Partanen, J; Haimila, K; Wieser, H; Mäki, M; Kaukinen, K. Secretion of celiac disease autoantibodies after in vitro gliadin challenge is dependent on small-bowel mucosal transglutaminase 2-specific IgA deposits.*BMC Immunol*. 2008 9, 6.

[13] Ron, Shaoul & Aaron, Lerner. Associated autoantibodies in celiac disease. *Autoimmunity reviews*. 2007 6, 559-565.

[14] Utz, PJ & Anderson P. Posttranslational protein modifications, apoptosis, and the bypass of tolerance to autoantigens *Arthritis Rheum*. 1998 41, 1152-1160.

[15] Arrieta, MC; Bistritz, L; Meddings, JB. Alterations in intestinal permeability. *Gut*. 2006 55, 1512-1520.

[16] Szebeni, B; Veres, G; Dezsofi, A; Rusai, K; Vannay, A; Bokodi, G; Vásárhelyi, B; Korponay-Szabó, IR; Tulassay, T; Arató, A. Increased mucosal expression of Toll-like receptor (TLR)2 and TLR4 in coeliac disease. *J Pediatr Gastroenterol Nutr* 2007 45, 187–193.

[17] Hue, S; Mention, JJ; Monteiro, RC; Zhang, S; Cellier, C; Schmitz, J; Verkarre, V; Fodil, N; Bahram, S; Cerf-Bensussan, N; Caillat-Zucman, S. A direct role for NKG2D/MICA interaction in villous atrophy during celiac disease. *Immunity*. 2004 21,367-377.

[18] Meresse, B; Chen, Z; Ciszewski, C; Tretiakova, M; Bhagat, G; Krausz, T; Raulet, D; Lanier, T; Groh, V; Spies, T. Coordinated Induction by IL15 of a TCR-Independent NKG2D Signaling Pathway Converts CTL into Lymphokine-Activated Killer Cells in Celiac Disease. *Immunity* 2004 21, 357–366.

[19] Schumanm, M; Richter, JF; Wedell, I; Moos, V; Zimmermann-Kordmann, M; Schneider, T; Daum, S; Zeitz, M; Fromm, M; Schulzke, JD. Mechanisms of epithelial translocation of the alpha(2)-gliadin-33mer in coeliac sprue. *Gut* 2008 57, 747-754.

[20] Orrù, S; Caputo, I; D'Amato, A; Ruoppolo, M; Esposito, C. Proteomics identification of acyl-acceptor and acyl-donor substrates for transglutaminase in a human intestinal epithelial cell line. Implications for celiac disease. *J Biol Chem*. 2003 278, 31766-31773.

[21] Piredda, L; Amendola, A; Colizzi, V; Davies, PJ; Farrace, MG; Fraziano, M; Gentile, V; Uray, I; Piacentini, M; Fesus, L. Lack of 'tissue' transglutaminase protein cross-linking leads to leakage of macromolecules from dying cells, relationship to development of autoimmunity in MRLIpr/Ipr mice. *Cell Death Differ*. 1997 4, 463-72.

[22] Nicholas, B; Smethurst, P; Verderio, E; Jones, R; Griffin, M. Cross-linking of cellular proteins by tissue transglutaminase during necrotic cell death, a mechanism for maintaining tissue integrity. *Biochem J.* 2003 371, 413-22.

[23] Carroccio, A; Brusca, I; Iacono, G; Di Prima, L; Teresi, S; Pirrone, G; Florena, AM; La Chiusa, SM; Averna, MR. Correlation with Intestinal Mucosa Damage and Comparison of ELISA with the Immunofluorescence Assay. *Clin. Chem*. 2005 51, 917- 920.

[24] Multhoff G. Heat shock proteins in immunity. *Handb Exp Pharmacol.* 2006 172, 279-304.

[25] Caggiari, L; Cannizzaro, R; De Zorzi, M; Canzonieri, V; Da Ponte, A; De Re V. A new HLA-A*680106 allele identified in individuals with celiac disease from the Friuli area of northeast Italy. *Tissue Antigens*. 2008 Epub ahead of print.

[26] Collin, P; Kaukinen, K; Vogelsang, H; Korponay-Szabo, I; Sommer, R; Schreier, E; Volta, U; Granito, A; Veronesi, L; Mascart, F; Ocmant, A; Ivarsson, A; Lagerqvist, C; Burgin-Wolff, A; Hadziselimovic, F; Furlano, RI; Sidler, MA; Mulder, CJ; Goerres, MS; Mearin, ML; Ninaber, MK; Gudmand-Hoyer, E; Fabiani, E; Catassi, C; Tidlund, H; Alainentalo, L; Maki, M. Antiendomysial and antihuman recombinant tissue transglutaminase antibodies in the diagnosis of coeliac disease, a

biopsy-proven European multicentre study. *Eur J Gastroenterol Hepatol* 2005 17, 85–91.

[27] Salmi, T; Collin, P; Korponay-Szabó, I; Laurila, K; Partanen, J; Huhtala, H; Király, R; Lorand, L; Reunala, T; Mäki, M; Kaukinen, K. Endomysial antibody-negative coeliac disease, clinical characteristics and intestinal autoantibody deposits. *Gut* 2006 55, 1746–1753.

[28] Stulík, J; Hernychová, L; Porkertová, S; Pozler, O; Tucková, L; Sánchez, D; Bures, J. Identification of new celiac disease autoantigens using proteomic analysis. *Proteomics*. 2003 3, 951-956.

[29] Maiero, S; Simula, M.P; Canzonieri, V; Marin, M.D; De Zorzi, M; Caggiari, L; Cannizzaro, R; De Re V. PPAR signalling pathway is involved in CD-associated inflammation. *Digestive and Liver Disease*. 2008 40, S9-S9.

[30] Barabási, AL & Oltvai, ZN. Network biology, understanding the cell's functional organization. *Nat Rev Genet*. 2004 5, 101-113.

[31] Bünger, M; van den Bosch, HM; van der Meijde, J; Kersten, S; Hooiveld, GJ; Müller, M. Genome-wide analysis of PPARalpha activation in murine small intestine. *Physiol Genomics*. 2007 30: 192-204.

[32] Tong-Chuan, HE; Chan, TA; Vogelstein, B; Kinzler' KW. PPAR delta is an APC-regulated target of nonsteroidal anti-inflammatory drugs. *Cell* 1999 99, 335–345.

[33] Lefebrve, A-M; Najib, J; Dewreumaux, P; Najib, J; Fruchart, JC; Geboes, K; Briggs, M; Heyman, R; Auwerx, J. Activation of the peroxisome proliferator activated receptor gamma promotes the development of colon tumours in C57BL/6J-APC^{Min}/+ mice. *Nat Med* 1998 4, 1053–1057.

[34] Saez, E; Tontonoz, P; Nelson, MC; Alvarez, JG; Ming, UT; Baird, SM; Thomazy, VA; Evans, RM. Activation of the nuclear receptor PPAR gamma enhance colon polyp formation. *Nat Med* 1998 4, 1058–1061.

[35] Curley, CR; Monsuur, AJ; Wapenaar, MC; Rioux, JD; Wijmenga, C. A functional candidate screen for coeliac disease genes. *Eur J Hum Genet*. 2006 14, 1215-1222.

[36] Curley, CR; Monsuur, AJ; Wapenaar, MC; Rioux, JD, Wijmenga, C. A functional candidate screen for coeliac disease genes, *Eur J Hum Genet* 2006 14, 1215–1222.

[37] Crabtree, JE; Heatley, RV; Losowsky, ML. Immunoglobulin secretion by isolated intestinal lymphocytes, spontaneous production and T cell

regulation in normal small intestine and in patients with coeliac disease. *Gut* 1989 30, 347–354.

[38] Scott, BB; Scott, DG; Losowsky, MS. Jejunal mucosal immunoglobulins and complement in untreated coeliac disease. *J Pathol.* 1977 121, 219-223.

[39] Halstensen, TS; Hvatum, M; Scott, H; Fausa, O; Brandtzaeg, P. Association of subepithelial deposition of activated complement and immunoglobulin G and M response to gluten in celiac disease. *Gastroenterology*. 1992 102, 751-759.

In: Proteomics
Editor: Giselle C. Rancourt
ISBN: 978-1-61668-691-8

Chapter 2

The Puzzle of Protein Location in Plant Proteomics

Elisabeth Jamet and Rafael Pont-Lezica
Surfaces Cellulaires et Signalisation chez les Végétaux,
Université de Toulouse, Pôle de Biotechnologie Végétale,
Castanet-Tolosan, France.

Abstract

Organelle proteomics allows the characterization of complex proteomes to understand the protein networks which regulate growth and development, as well as adaptation and evolution. Purification of organelles is of paramount importance and diverse protocols are published. Some organelles such as chloroplasts, mitochondria, and the nucleus are surrounded by membranes which facilitate their purification. Others have membranes easily disrupted (vacuoles and peroxisomes), or are complex systems for protein trafficking (endoplasmic reticulum, Golgi, and secretory vesicles). The cell walls present different difficulties since they have no physical limits allowing purification. The purity of the targeted cell compartment is usually evaluated by biochemical and/or immunological methods. Nevertheless, in any sub-cellular proteomic analysis, proteins from a different compartment can be detected and the difficulty is to decide whether it is a contamination, or the unexpected location is real and has a functional significance. Software to predict sub-cellular location of proteins is available. However, since not all the targeting signals are known at present, carefulness in the use of such tools is recommended. Different tactics to solve this puzzle are discussed in this commentary.

INTRODUCTION

Plants like other organisms depend on proteins to maintain their functions and to adjust to environmental changes. Proteins arrange themselves into metabolic and regulatory pathways and their accurate localization is essential for the organism. To understand the diverse protein functions, information on the identification of all the proteins, as well as the knowledge of their location, post-translational modifications, and quantitative changes in the cell are important. Proteomics can provide the basic information, but the main challenge is the biochemical complexity, and the dynamic range of proteins. Since the cell is structured in organelles with different but interconnected functions, proteomic analysis of purified cell organelles reduces sample complexity to a subset of functionally related proteins or pathway modules [51, 56]. Reliable protein localization by proteomics requires that either organelle preparations are free of contaminants or that techniques are used to discriminate between genuine organelle residents and contaminating proteins [12, 23, 34]. Nonetheless, plant organelle proteomics has been restricted by the difficulties in isolating pure sub-cellular compartments in sufficient amount.

ORGANELLE ISOLATION

Organelle isolation implies the choice of an appropriate method for the purification of the targeted organelle. Reasonably pure preparations of some organelles surrounded by a double membrane, such as chloroplasts and mitochondria, can be achieved [20, 27, 33]. Cells may be released from tissues by enzymatic treatment or mechanical disruption; the last one being the most frequently used in plant proteomics. The exposure time as well as the strength of the forces applied has to be optimized to avoid a progressive destruction of the biological supramolecular architecture; otherwise large-scale destruction of organelles will occur and organelle yield is compromised. Nevertheless, the purity of organelle preparations can be established using microscopic observations, marker enzymes or antibodies against known proteins. The difficulty is that the analysis tool (mass spectrometry) is 100 to 1000 times more sensitive than classical biochemical and immunological tests [25]. In certain cases, microscopic observations proved to be helpful to check the quality of a sample [5, 6, 23], but they can be subjective. It is then hazardous to use results of sub-cellular proteomics directly as a proof for the location of the identified proteins.

ORGANELLES SURROUNDED BY DOUBLE MEMBRANES

From the protein composition point of view, mitochondria and chloroplasts are quite complex and include both very soluble proteins (present in the matrix and the intermembrane space) and very hydrophobic membrane proteins. In addition, there are membrane proteins of intermediate solubility, such as some subunits of the oxidative phosphorylation complexes or the outer membrane porins. This chemical heterogeneity is a real challenge for the proteomic analysis.

Most of the chloroplast proteomic studies started with the *Arabidopsis* genome sequencing. Predictions indicate that 2100 to 2700 proteins are located in the *Arabidopsis* chloroplast [56, 57]. However, only about half of these proteins have been experimentally identified [22]. The purification of intact chloroplasts by percoll gradient centrifugation produces reasonably pure preparations. However, chloroplasts are complex organelles composed of six distinct suborganellar compartments: three different membranes (the two envelope membranes and the internal thylakoid membrane) and three discrete aqueous compartments (the intermembrane space of the envelope, the stroma and the thylakoid lumen). As a result of this structural intricacy, the external and internal routing of chloroplast proteins is necessarily a complex process [27, 28]. Proteomics of membrane proteins is challenging since they are poorly resolved using classical 2D-electrophoresis techniques [53, 66]. Proteins predicted to be targeted to the endoplasmic reticulum (ER) and mitochondria have been identified in chloroplast proteomic studies [35, 62], pointing to the need for experimental determination of protein location and the relative value of proteomic analysis.

Mitochondria can be isolated from many tissues but *Arabidopsis* cell suspension cultures have been largely used. On average, samples of 90-95 % purity were obtained using a Percoll gradient method [61], and yeast mitochondrial preparations reached 98% purity using free-flow electrophoresis [65]. About one fourth of the expected 2000-2500 *Arabidopsis* mitochondrial proteins were identified by proteomics [39]. The sub fractionation of mitochondria into the four basic compartments (outer membrane, inter-membrane space, inner membrane, and matrix) involves a complex protocol. As for chloroplasts, the proteomic characterization of mitochondrial hydrophobic proteins is a difficult task because of their low abundance and insolubility [8, 41, 53]. Several pathways are presently explored to solve the problem of contamination by non-mitochondrial proteins [6, 39].

As other organelles delimited by two membranes, nuclei isolation and purification were performed on different plants by density gradient centrifugation [2, 9, 32]. Most of the proteins identified were verified nuclear proteins, but also non-classical nuclear proteins were identified in different plants. A comparative study of the nuclear proteomes of *Arabidopsis*, rice and chickpea was done; from the 382 proteins identified in the three proteomes, only a small proportion were orthologous proteins, the others being specific for each plant [9]. Most of the nuclear proteins show a high level of divergence in the protein classes between the three plants, which is surprising for an organelle having similar functions in different organisms.

Other Organelles

The isolation of other organelles such as peroxisomes [1], vacuoles [26, 48] and oil bodies [30] produced samples enriched in the selected organelle. The purity of the selected organelle can be evaluated by western blotting using antibodies specific against proteins from undesirable compartments [26]. In these preparations, the proportion of proteins known to be present in other compartments depends on the purification method. Some authors pretend that those are new proteins for the studied compartment, others just call those proteins contaminants. However it is clear that the location of such proteins should be verified by other methods than proteomics.

Although reasonably pure preparations of some organelles (chloroplasts and mitochondria) can be achieved by centrifugation and density gradients, the isolation of other compartments can not be accomplished by centrifugation because they do not have specific buoyant densities. Such is the case of the endomembrane system which has proven recalcitrant to purification [19, 49]. An additional problem with the endomembrane system is that proteins move along the different organelles in a controlled way. Some are residents of a particular compartment such as the ER or the Golgi, and others are sent to specific compartments (plasma membrane, tonoplast, and vacuole) or are secreted outside the cell. A well-designed technique was developed for such organelles employing differential isotope labelling: LOPIT (Localization of Organelle Proteins by Isotope Tagging) [11, 13]. LOPIT is based on the partial separation of organelles by density gradient centrifugation followed by the analysis of protein distributions within the gradient by Isotope-Coded Affinity Tag (ICAT) labeling and mass spectrometry (MS). This technique does not depend on the production of pure organelles, and enables proteins from different sub-cellular compartments to be distinguished even if their distributions overlap.

THE CASE OF THE CELL WALL

The cell wall cannot be defined as an organelle, but it is a very important cell compartment which requires designing specific strategies for its purification [24]. The lack of a delimiting membrane may result in the lost of proteins during its purification and in contamination by other cell compartments. The polysaccharide networks constitute potential traps for intracellular proteins. Two types of strategies were tried: non-destructive and destructive ones [24, 38]. In both cases, proteins not expected to be present in cell walls were found, leading to the hypothesis of the existence of an alternative secretory pathway [60]. However, such a pathway, if it exists, probably targets only a few proteins to the cell wall. Indeed, intracellular proteins identified in cell wall proteomics vary from one experiment to another, and it is possible to reduce their number by maintaining the integrity of plasma membranes or improving the cell wall purification procedure [5, 6, 17].

Altogether, the results of sub-cellular proteomics contain a certain number of proteins for which a clear localization cannot be predicted. The relatively high number of proteins without a known function identified by proteomics also raises the question of their function and location. Proteomic results start to be incorporated into databases such as those listed in Table I. This will make easier comparisons between results obtained in different conditions, and both allow confirmation of sub-cellular localization and point at possible contaminants.

Table I. Plant proteomic databases

Database Name	Cell compartment	Web site	Species	Reference
PPDB	any compartment	http://ppdb.tc.cornell.edu/	*Arabidopsis thaliana, Zea mays*	18 Friso *et al.* 2004
SUBA	any compartment	http://www.plantenergy.uwa.edu.au/suba2/	*Arabidopsis thaliana*	22 Heazlewood *et al.* 2007
PRIDE	any compartment	http://www.ebi.ac.uk/pride/	all organisms	31 Jones *et al.* 2008
NASC proteomics database	any compartment	http://proteomics.arabidopsis.info/	*Arabidopsis thaliana*	
AtProteome	any compartment	http://fgcz-atproteome.unizh.ch/	*Arabidopsis thaliana*	3 Baerenfaller *et al.* 2008
eSLDB	any compartment	http://gpcr2.biocomp.unibo.it/esldb/index.htm	all organisms	50 Pierleoni et al. 2007
WallProtDB	cell wall	http://www.polebio.scsv.ups-tlse.fr/WallProtDB/	*Arabidopsis thaliana, Oryza sativa*	52 Pont-Lezica *et al.* 2009
PlantProteomics07	plasma membrane	http://www.grenoble.prabi.fr/data/PlantProteomics07/	*Arabidopsis thaliana*	16 Ephiritikhine *et al.* 2008
plProt	plastid	http://www.plprot.ethz.ch/	*Arabidopsis thaliana, Nicotiana tabacum*	35 Kleffmann *et al.* 2006
AMPDB	mitochondrion	http://www.plantenergy.uwa.edu.au/applications/ampdb/	*Arabidopsis thaliana*	21 Heazlewood *et al.* 2005
AMPP	mitochondrion	http://www.gartenbau.uni-hannover.de/arabidopsis.html	*Arabidopsis thaliana*	36 Kruft *et al.* 2001
AtNoPDB	nucleolus	http://bioinf.scri.sari.ac.uk/cgi-bin/atnopdb/home	*Arabidopsis thaliana*	7 Brown *et al.*, 2005

PROTEIN LOCATION

Most proteins are synthesized on cytosolic ribosomes, except for chloroplasts and mitochondria having their own synthesis capacities in addition to the nuclear-encoded proteins. After synthesis, proteins can be post-translationally modified, leading to differential location. The accurate protein location is essential for all living organisms and organelle proteomic analysis can generate very large candidate list of putative constituent proteins. Therefore independent approaches are required to verify if the candidate proteins can be included in the targeted organelle. To establish the location profiles of large set of proteins, new tools were developed that generated high-throughput localization data of a large number of proteins.

BIOINFORMATIC PREDICTIONS

Many proteins may contain within their amino acid sequence information that could be used for predicting their sub-cellular location, *e.g.* signal peptides, transit peptides, nuclear import and export signals [64]. Numerous computational programs have been developed with the aim of predicting the location of the proteins, most of which rely on the presence of these signals [15, 37, 45, 47]. It is important to understand the uses and limits of the bioinformatic methods developed lately, and recent reviews give a comprehensive approach on the available tools [14, 34].

Table II. Bioinformatic tools for plant sub-cellular proteomics

Database name	Cell compartment	Web site	Species	Reference
ProtAnnDB	any compartment	http://www.polebio.scsv.ups-tlse.fr/ProtAnnDB/	*Arabidopsis thaliana*, *Oryza sativa*	58 San-Clemente *et al.* 2009
Aramemnon	membranes	http://aramemnon.botanik.uni-koeln.de/	*Arabidopsis thaliana*, *Oryza sativa*, *Vitis vinifera*, *Zea mays*, *Other seed plants*	59 Schwacke *et al.* 2003

Since computational methods are based on different algorithms, it is recommended to use several prediction programs to make a reasonable choice. Two tools available on line are listed in Table II. Both Aramemnon and *ProtAnnDB* use a series of available software predicting sub-cellular localization of proteins and compare their results [58, 59]. In addition, recent reports suggest the existence of alternative sorting routes for proteins. For instance, the chloroplastic ceQORH protein was shown to be synthesized without a canonical cleavable chloroplastic transit sequence [42, 43]. Other proteins were shown to be targeted to the chloroplast *via* the secretory pathway and undergo glycosylation [46, 54, 63]. Moonlighting proteins make the interpretation of proteomic results more difficult since those proteins are located in several cell compartments where they have different functions [29].

Proteins routed through the secretory pathway can be predicted using different tools [14]. No specific prediction program exists for vacuole, and therefore, localization can only be inferred *via* experimental data or homology searching referring to well-annotated proteins. Despite the knowledge of alternative secretion pathways, no current machine learning approach directly addresses the problem of predicting proteins entering the non-classical secretory pathway. However, prediction methods based on amino acid composition are in principle capable of foretelling proteins entering the non-classical secretory pathway [55]. A sequence-based method of prediction of non-classical-triggered secretion for mammalian and bacterial proteins has also been developed [4]. The assumption is that extracellular proteins share certain properties and features which can be related to protein function outside the cell, independently of the secretory process itself. Fifty proteins found in several proteomic studies of *Arabidopsis* cell walls and devoid of predicted signal peptide were analyzed. Only 14 were predicted as putative secreted proteins [24], suggesting that this alternative pathway is highly exceptional.

EXPERIMENTAL LOCATION

The experimental testing of protein targeting is the best way to verify the computational predictions. Several methods have been used and their accuracy is uneven. The first protein localizations were made by biochemical methods, and some abundant proteins were characterized and turned into marker enzymes for several cell compartments. The biochemical tools consist mainly in destructive approaches allowing cell fractionation. Fractionation has several restrictions since, as mentioned above, some compartments cannot be easily separated; still those organelles that can be fractionated are not free of contaminants [23]. Immunoprecipitation of tagged proteins may be useful to identify other proteins within a complex.

Cytological localization is a non-destructive technique allowing a good localization of proteins. Immunolabeling of cells supply a good resolution and high specificity, depending on the antibody. Proteins encoded by multigenic families may be difficult to localize precisely because highly specific antibodies are necessary to discriminate them. This is critical when different members of such protein families are targeted to different compartments. The development of a small peptide tag for covalently label proteins such as *c*-myc or His is a good approach because a larger protein tags may affect the function of the protein of interest. This allows the use of commercial antibodies.

One of the most popular methods to localize proteins *in-vivo* is the Green Fluorescent Protein (GFP)-fusion technology. GFP is a small protein from jellyfish that can be visualized by fluorescence in a non-invasive way [10]. *In-vivo* uptake assays are widespread since they maintain the cellular environment, and targeting into all organelles can be tested at the same time. The principal drawback is that the tag may modify the location of the native protein and may even alter the condition of the wider system through dominant effects. It is important to test such constructs in a null mutant for the studied protein and with the native expression signals; if not there is inevitably a degree of overexpression. Another important point is to verify that the distribution of the GFP signal is consistent with gene expression visualized via *in situ* RNA hybridization. Localization of proteins in the endomembrane system is more complex than in chloroplasts or mitochondria since there is an exchange of membrane constituents. Furthermore the proteins of interest may be distributed between more than one compartment, and still cycle between them [40, 44]. The critical point is the subjective interpretation one puts on each observation in light of the experimental conditions and other pertinent data.

CONCLUSION

The results of sub-cellular proteomics cannot be considered sufficient to ensure the correct determination of protein location in cells. The use of several localization techniques is recommended, even if this seems to be redundant.

Bioinformatic software based on experimental data provides valuable tools to predict sub-cellular localization of proteins. Even if a protein can accumulate at several places in the cell and be active at a place where it does not accumulate, combining approaches that take into account targeting and accumulation of proteins may give more confident localization [40]. Finally, the integration of localization data with expected biological function and within metabolic networks is important to confirm the location of a particular protein.

Acknowledgments

The authors are thankful to CNRS and Université Paul Sabatier (Toulouse III, France) for supporting their research.

References

[1] Arai, Y., Youichiro, F., Hayashi, M. and Nishimura, M. (2008). Peroxisome, in Agrawal, G. and Rakwal, R., Editors. *Plant Proteomics: Technologies, Strategies, and Applications*. J. Wiley & Sons: Hoboken, NJ. pp. 377-389.

[2] Bae, M. S., Cho, E. J., Choi, E. Y. and Park, O. K. (2003). Analysis of the *Arabidopsis* nuclear proteome and its response to cold stress. *Plant J. 36*: 652-663.

[3] Baerenfaller, K., Grossmann, J., Grobei, M. A., Hull, R., Hirsch-Hoffmann, M., Yalovsky, S., Zimmermann, P., Grossniklaus, U., Gruissem, W. and Baginsky, S. (2008). Genome-scale proteomics reveals *Arabidopsis thaliana* gene models and proteome dynamics. *Science 320*: 938-941.

[4] Bendtsen, J. D., Jensen, L. J., Blom, N., Von Heijne, G. and Brunak, S. (2004). Feature-based prediction of non-classical and leaderless protein secretion. *Protein Eng. Des. Sel. 17*: 349-356.

[5] Borderies, G., Jamet, E., Lafitte, C., Rossignol, M., Jauneau, A., Boudart, G., Monsarrat, B., Esquerré-Tugayé, M. T., Boudet, A. and Pont-Lezica, R. (2003). Proteomics of loosely bound cell wall proteins of *Arabidopsis thaliana* cell suspension cultures: A critical analysis. *Electrophoresis 24*: 3421-3432.

[6] Boudart, G., Jamet, E., Rossignol, M., Lafitte, C., Borderies, G., Jauneau, A., Esquerré-Tugayé, M.-T. and Pont-Lezica, R. (2005). Cell wall proteins in apoplastic fluids of *Arabidopsis thaliana* rosettes: Identification by mass spectrometry and bioinformatics. *Proteomics 5*: 212-221.

[7] Brown, J. W., Shaw, P. J., Shaw, P. and Marshall, D. F. (2005). Arabidopsis nucleolar protein database (AtNoPDB). *Nucleic Acids Res. 33*: D633-636.

[8] Brugière, S., Kowalski, S., Ferro, M., Seigneurin-Berny, D., Miras, S., Salvi, D., Ravanel, S., d'Herin, P., Garin, J., Bourguignon, J., Joyard, J. and Rolland, N. (2004). The hydrophobic proteome of mitochondrial membranes from *Arabidopsis* cell suspensions. *Phytochemistry 65*: 1693-1707.

[9] Chakraborty, S., Pandey, A., Datta, A. and Chakraborty, N. (2008). Nucleus, in Agrawal, G. and Rakwal, R., Editors, *Plant Proteomics: Technologies, Strategies, and Applications*, J. Wiley & Sons Hoboken, NJ. pp. 327-338.

[10] Chalfie, M. (2009). GFP: Lighting up life. Proc. Natl. Acad. Sci. *U S A 106*: 10073-10080.

[11] Dunkley, T., Hester, S., Shadforth, I., Runions, J., Weimar, T., Hanton, S., Griffin, J., Bessant, C., Brandizzi, F., Hawes, C., Watson, R., Dupree, P. and Lilley, K. (2006). Mapping the *Arabidopsis* organelle proteome. *Proc. Natl. Acad. Sci. U S A 103*: 6518-6523.

[12] Dunkley, T., Watson, R., Griffin, J., Dupree, P. and Lilley, K. (2004). Localization of organelle proteins by isotope tagging (LOPIT). *Mol. Cell. Proteomics 3*: 1128-1134.

[13] Dunkley, T. P., Dupree, P., Watson, R. B., and Lilley, K. S. (2004). The use of isotope-coded affinity tags (ICAT) to study organelle proteomes in *Arabidopsis thaliana. Biochem. Soc. Trans. 32*: 520-523.

[14] Emanuelsson, O., Brunak, S., von Heijne, G. and Nielsen, H. (2007). Locating proteins in the cell using TargetP, SignalP and related tools. *Nat. Protoc. 2*: 953-971.

[15] Emanuelsson, O., Nielsen, H., Brunak, S. and von Heijne, G. (2000). Predicting subcellular localization of proteins based on their N-terminal amino acid sequence. *J. Mol. Biol. 300*: 1005-1016.

[16] Ephiritikhine, G., Marmagne, A., Meinnel, T. and Ferro, M. (2008). Plasma membrane: A peculiar status among the cell membrane systems, in Agrawal, G. and Rakwal, R., Editors, *Plant Proteomics:*

Technologies, Strategies, and Applications. J. Wiley & Sons: Hoboken, NJ. pp. 309-326.

[17] Feiz, L., Irshad, M., Pont-Lezica, R. F. and Canut, H. (2006). Evaluation of cell wall preparations for proteomics: a new procedure for purifying cell walls from *Arabidopsis* hypocotyls. *Plant Methods 2*: 10.

[18] Friso, G., Giacomelli, L., Ytterberg, A., Peltier, J., Rudella, A., Sun, Q. and van Wijk, K. (2004). In-depth analysis of the thylakoid membrane proteome of *Arabidopsis thaliana* chloroplasts: New proteins, new functions, and a plastid proteome database. *Plant Cell 16*: 478-499.

[19] Gabaldon, T. and Huynen, M. A. (2004). Shaping the mitochondrial proteome. *Biochim. Biophys. Acta 1659*: 212-220.

[20] Heazlewood, J. and Millar, A. (2007). *Arabidopsis* mitochondrial proteomics. *Methods Mol. Biol. 372*: 559-571.

[21] Heazlewood, J. L. and Millar, A. H. (2005). AMPDB: the *Arabidopsis* Mitochondrial Protein Database. *Nucleic Acids Res. 33*: D605-610.

[22] Heazlewood, J. L., Verboom, R. E., Tonti-Filippini, J., Small, I. and Millar, A. H. (2007). SUBA: the *Arabidopsis* Subcellular Database. *Nucleic Acids Res. 35*: D213-218.

[23] Huber, L., Pfaller, K. and Vietor, I. (2003). Organelle proteomics: implications for subcellular fractionation in proteomics. *Circ. Res. 92*: 962-968.

[24] Jamet, E., Canut, H., Albenne, C., Boudart, G. and Pont-Lezica, R. (2008). Cell Wall, in Agrawal, G. and Rakwal, R., Editors, *Plant Proteomics: Technologies, Strategies, and Applications*. J. Wiley & Sons: Hoboken, NJ. pp. 293-307.

[25] Jamet, E., Canut, H., Boudart, G. and Pont-Lezica, R. F. (2006). Cell wall proteins: a new insight through proteomics. *Trends Plant Sci. 11*: 33-39.

[26] Jaquinod, M., Villiers, F., Kieffer-Jaquinod, S., Hugouvieux, V., Bruley, C., Garin, J. and Bourguignon, J. (2007). A proteomics dissection of *Arabidopsis thaliana* vacuoles from cell culture. *Mol. Cell Proteomics 6*: 394-412.

[27] Jarvis, P. (2007). The proteome of chloroplasts and other plastids, in Samaj, J. and Thelen, J., Editors, *Plant Proteomics.* Springer-Verlag: Berlin, Heidelberg. pp. 207-225.

[28] Jarvis, P. (2008). Targeting of nucleus-encoded proteins to chloroplasts in plants. *New Phytol. 179*: 257-285.

[29] Jeffery, C. J. (2005). Mass spectrometry and the search for moonlighting proteins. *Mass Spectrom. Rev. 24*: 772-782.

[30] Jolivet, P., Negroni, L., d'Andréa, S. and Chardot, T. (2008). Oil bodies, in Agrawal, G. and Rakwal, R., Editors, *Plant Proteomics: Technologies, Strategies, and Applications*. J. Wiley & Sons: Hoboken, NJ., pp. 407-417.

[31] Jones, P., Côté, R., Cho, S. Y., Klie, S., Martens, L., Quinn, A., Thorneycroft, D. and Hermjakob, H. (2008). PRIDE: new developments and new datasets. *Nucleic Acids Res. 36*: D878-883.

[32] Khan, M. M. and Komatsu, S. (2004). Rice proteomics: recent developments and analysis of nuclear proteins. *Phytochemistry 65*: 1671-1681.

[33] Kieselbach, T. and Schröder, W. (2008). Chloroplast, in Agrawal, G. and Rakwal, R., Editors, *Plant Proteomics: Technologies, Strategies, and Applications*. J. Wiley & Sons: Hoboken, NJ. pp. 339-350.

[34] Kitsios, G., Tsesmetzis, N., Bush, M. and Doonan, J. (2008). The *Arabidopsis* localizome: Subcellular protein localization and interactions in *Arabidopsis*, in Agrawal, G. and Rakwal, R., Editors, *Plant Proteomics: Technologies, Strategies, and Applications*. J. Wiley & Sons: Hoboken, NJ. pp. 61-81.

[35] Kleffmann, T., Hirsch-Hoffmann, M., Gruissem, W. and Baginsky, S. (2006). plprot: a comprehensive proteome database for different plastid types. *Plant Cell Physiol. 47*: 432-436.

[36] Kruft, V., Eubel, H., Jansch, L., Werhahn, W. and Braun, H. P. (2001). Proteomic approach to identify novel mitochondrial proteins in *Arabidopsis*. *Plant Physiol. 127*: 1694-1710.

[37] la Cour, T., Gupta, R., Rapacki, K., Skriver, K., Poulsen, F. M. and Brunak, S. (2003). NESbase version 1.0: a database of nuclear export signals. *Nucleic Acids Res. 31*: 393-396.

[38] Lee, S. J., Saravanan, R. S., Damasceno, C. M., Yamane, H., Kim, B. D. and Rose, J. K. (2004). Digging deeper into the plant cell wall proteome. *Plant Physiol. Biochem. 42*: 979-988.

[39] Millar, A. (2007). The plant mitochondrial proteome, in Samaj, J. and Thelen, J., Editors, *Plant proteomics*. Springer-Verlag: Berlin Heidelberg. pp. 226-246.

[40] Millar, A., Chris Carrie, C., Pogson, B. and Whelan, J. (2009). Exploring the function-location nexus: Using multiple lines of evidence in defining the subcellular location of plant proteins. *Plant Cell 21*: 1625-1631.

[41] Millar, A. H. and Heazlewood, J. L. (2003). Genomic and proteomic analysis of mitochondrial carrier proteins in *Arabidopsis*. *Plant Physiol. 131*: 443-453.

[42] Miras, M., Salvi, D., Piette, L., Seigneurin-Berny, D., Grunwald, D., Reinbothe, C., Joyard, J., Reinbothe, S. and Rolland, N. (2007). TOC159- and TOC75-independent import of a transit sequence less precursor into the inner envelope of chloroplasts. *J. Biol. Chem. 282*: 29482-29492.

[43] Miras, S., Salvi, D., Ferro, M., Grunwald, D., Garin, J., Joyard, J. and Rolland, N. (2002). Non-canonical transit peptide for import into the chloroplast. *J. Biol. Chem. 277*: 47770-47778.

[44] Moore, I. and Murphy, A. (2009). Validating the location of fluorescent protein fusions in the endomembrane system. *Plant Cell 21*: 1632-1636.

[45] Nakai, K. and Horton, P. (1999). PSORT: a program for detecting sorting signals in proteins and predicting their subcellular localization. *Trends Biochem. Sci. 24*: 34-36.

[46] Nanjo, Y., Oka, H., Ikarashi, N., Kaneko, K., Kitajima, A., Mitsui, T., Muñoz, F., Rodríguez-López, M., Baroja-Fernández, E. and Pozueta-Romero, J. (2006). Rice plastidial N-glycosylated nucleotide pyrophosphatase/phosphodiesterase is transported from the ER-golgi to the chloroplast through the secretory pathway. *Plant Cell Physiol. 18*: 2582-2592.

[47] Nielsen, H., Engelbrecht, J., Brunak, S. and von Heijne, G. (1997). Identification of prokaryotic and eukaryotic signal peptides and prediction of their cleavage sites. *Prot. Eng. 10*: 1-6.

[48] Pan, S. and Raikhel, N. (2008). Unraveling plant vacuoles by proteomics, in Agrawal, G. and Rakwal, R., Editors, *Plant Proteomics: Technologies, Strategies, and Applications*. J. Wiley & Sons: Hoboken, NJ. pp. 391-405.

[49] Peck, S. C. (2005). Update on proteomics in *Arabidopsis*. Where do we go from here? *Plant Physiol. 138*: 591-599.

[50] Pierleoni, A., Martelli, P. L., Fariselli, P. and Casadio, R. (2007). eSLDB: eukaryotic subcellular localization database. *Nucleic Acids Res. 35*: D208-212.

[51] Ploscher, M., Granvogl, B., Reisinger, V., Masanek, A. and Eichacker, L. (2009). Organelle proteomics. *Methods Mol. Biol. 519*: 65-82.

[52] Pont-Lezica, R., Minic, Z., Roujol, D., San Clemente, H. and Jamet, E. (2009). Plant cell wall functional genomics: Novelties from proteomics, in Columbus, F., Editor, *Plant Genomics: Processes, Methods and Application*. Nova Science Publishers: Hauppauge, NY. (in press).

[53] Rabilloud, T. (2009). Membrane proteins and proteomics: love is possible, but so difficult. *Electrophoresis 30*: S174-180.

[54] Radhamony, R. and Theg, S. (2006). Evidence for an ER to Golgi to chloroplast protein transport pathway. *Trends Cell Biol. 16*: 385-387.

[55] Reinhardt, A. and Hubbard, T. (1998). Using neural networks for prediction of the subcellular location of proteins. *Nucleic Acids Res. 26*: 2230-2236.

[56] Reisinger, V. and Eichacker, L. (2009). Subcellular proteomics organelle proteomics: Reduction of sample complexity by enzymatic in-gel selection of native proteins. *Methods Mol. Biol. 564*: 325-333.

[57] Richly, E. and Leister, D. (2004). An improved prediction of chloroplast proteins reveals diversities and commonalities in the chloroplast proteomes of *Arabidopsis* and rice. *Gene 329*: 11-16.

[58] San Clemente, H., Pont-Lezica, R. and Jamet, E. (2009). Bioinformatics as a tool for assessing the quality of sub-cellular proteomics strategies and inferring functions of proteins: Plant cell wall proteomics as a test case. *Bioinform. Biol. Insights 3*: 15-28.

[59] Schwacke, R., Schneider, A., van der Graaff, E., Fischer, K., Catoni, E., Desimone, M., Frommer, W., Flügge, U. and Kunze, R. (2003). ARAMEMNON, a novel database for Arabidopsis integral membrane proteins. *Plant Physiol. 131*: 16-26.

[60] Slabas, A. R., Ndimba, B., Simon, W. J. and Chivasa, S. (2004). Proteomic analysis of the *Arabidopsis* cell wall reveals unexpected proteins with new cellular locations. *Biochem. Soc. Trans. 32*: 524-528.

[61] Tan, Y.-F. and Millar, H. (2008). The plant mitochondrial proteome and the challenge of hydrophobic protein analysis, in Agrawal, G. and Rakwal, R., Editors, *Plant Proteomics: Technologies, Strategies, and Applications*. J. Wiley & Sons Hoboken: NJ. pp. 361-376.

[62] van der Laan, M., Rissler, M. and Rehling, P. (2006). Mitochondrial preprotein translocases as dynamic molecular machines. *FEMS Yeast Res. 6*: 849-861.

[63] Villarejo, A., Burén, S., Larsson, S., Déjardin, A., Monné, M., Rudhe, C., Karlsson, J., Jansson, S., Lerouge, P., Rolland, N., von Heijne, G., Grebe, M., Bako, L. and Samuelsson, G. (2005). Evidence for a protein transported through the secretory pathway en route to the higher plant chloroplast. *Nat. Cell Biol. 7*: 1224-1231.

[64] von Heijne, G. (1990). Protein targeting signals. *Curr. Opin. Cell Biol. 2*: 604-608.

[65] Zischka, H., Weber, G., Weber, P. J., Posch, A., Braun, R. J., Buhringer, D., Schneider, U., Nissum, M., Meitinger, T., Ueffing, M. and Eckerskorn, C. (2003). Improved proteome analysis of *Saccharomyces*

cerevisiae mitochondria by free-flow electrophoresis. *Proteomics 3*: 906-916.

[66] Zychlinski, A. and Gruissem, W. (2009). Preparation and analysis of plant and plastid proteomes by 2DE. *Methods Mol. Biol. 519*: 205-220.

In: Proteomics
Editor: Giselle C. Rancourt

ISBN: 978-1-61668-691-8

Chapter 3

WHAT FUTURE FOR "GEL-BASED PROTEOMIC" APPROACHES?

François Chevalier*
Proteomic Laboratory, iRCM, CEA, Fontenay aux Roses, France

ABSTRACT

The simultaneous analysis of all proteins expressed by a cell, tissue or organism in a specific physiological condition is the main goal of proteomic studies. Gel-based proteomic is the most popular and versatile method of global protein separation and quantification. This is a mature approach to screen the protein expression at the large scale. Based on two independent biochemical characteristics of proteins, two-dimensional electrophoresis combines isoelectric focusing, which separates proteins according to their isoelectric point, and SDS-PAGE, which separates them further according to their molecular mass. The next typical steps of the flow of gel-based proteomics are spots visualization and evaluation, expression analysis and finally protein identification by mass spectrometry. At present, two-dimensional electrophoresis allows simultaneously to detect and quantify up to thousand protein spots in the same gel in a wide range of biological systems for the study of differentially expressed proteins. However, gel-based proteomic has a number of inherent drawbacks. In this review article, the benefits, difficulties, limits and perspectives of gel-based proteomic approaches are discussed.

---

* Tel: +33 (0)146 548 326, Fax: +33 (0)146 549 138, Email: francois.chevalier@cea.fr

Keywords: two-dimensional electrophoresis, proteomic methods, electrophoresis, isoelectric focusing, protein stain,

1 – Introduction

Proteomics, one of the most important areas of research in the post-genomic era, is not new in terms of its experimental foundations (1). It is a natural consequence of the huge advances in genome sequencing, bioinformatics and the development of robust, sensitive, reliable and reproducible analytical techniques (2-12). Genomics projects have produced a large number of DNA sequences from a wide range of organisms, including humans and mammals. This "genomics revolution" has changed the concept of the comprehensive analysis of biological processes and systems. It is now hypothesized that biological processes and systems can be described based on the comparison of global, quantitative gene expression patterns from cells or tissues representing different states. The discovery of posttranscriptional mechanisms that control rate of synthesis and half-life of proteins and the ensuing nonpredictive correlation between mRNA and protein levels expressed by a particular gene indicate that direct measurement of protein expression also is essential for the analysis of biological processes and systems. Global analysis of gene expression at the protein level is now also termed proteomics. The standard method for quantitative proteome analysis combines protein separation by high resolution (isoelectric focusing / SDS-PAGE) two-dimensional gel electrophoresis (2DE) with mass spectrometric (MS) or tandem MS (MS/MS) identification of selected protein spots (5, 9, 11, 13-16). Important technical advances related to 2DE and protein MS have increased sensitivity, reproducibility, and throughput of proteome analysis while creating an integrated technology. Quantitation of protein expression in a proteome provides the first clue into how the cell responds to changes in its surrounding environments. The resulting over- or under-expressed proteins are deemed to play important roles in the precise regulation of cellular activities that are directly related to a given exogenous stimulus. Conventional two-dimensional gel electrophoresis (2DE), in combination with advanced mass spectrometric techniques, has facilitated the rapid characterization of thousands of proteins in a single polyacrylamide gel. The uniqueness of 2DE for easy visualisation of protein isoforms, using two physical parameters such as isoelectric point and molecular weight, renders this technology itself extremely informative. The method routinely analyzes more than 1000 different protein spots separated on a single two-dimensional gel and, thus, is well suited for the global analysis of protein expression in an organism. However, high-throughput quantitation of proteins from different cell lysates remains a challenging issue, owing to the poor reproducibility of 2DE, as well as low sensitivity and narrow linear dynamic ranges in the detection methods (17-21). Recent developments of fluorescent dyes, such as the different commercially available SYPRO dyes, partially addressed some of these problems (22-30). These dyes detect as little as 1 ng of proteins, and at the same time they offer more than 1000-fold linear dynamic range. The more critical issue, however, is the reproducibility problem of 2DE. Even the identical protein samples that are run on two separate two-dimensional gels will normally produce very similar but not identical 2DE protein maps, owing to the gel-to-gel and operator-to-operator variations. This can be circumvented using multiplexing methods such as fluorescent two-dimensional "Difference Gel Electrophoresis" (2-D DIGE), which substantially reduces variability by displaying two or more complex protein mixtures labeled with different fluorescent dyes in a single 2D gel (21, 31-38).

In this review, we focus on the latest developments in 2DE within the context of large-scale proteomics to reveal the advantages, limits and perspectives of the 2DE-based proteomic approach.

2 - The State of the Art of "Gel-Based Proteomic" Approaches

2.1. Sample Preparation and Protein Solubilisation

In order to take advantage of the high resolution capacity of 2DE, proteins have to be completely denatured, disaggregated, reduced and solubilised to disrupt molecular interactions and to ensure that each spot represents an individual polypeptide.

Although a large number of standard protocols has been published, these protocols have to be adapted and further optimized for the type of sample (bacteria / yeast / mammalian cells; cells / tissue; animal / vegetal material; etc...) to be analyzed, as well as for the proteins of interest (cytosolic / nuclear; total "soluble" or membrane "insoluble" proteins; etc...).

After cell disruption, native proteins must be denatured and reduced to disrupt intra- and intermolecular interactions, and solubilized while maintaining the inherent charge properties. Sample solubilization is carried out using a buffer containing chaotropes (urea and/or thiourea), nonionic (Triton X-100) and/or zwitterionic detergents (CHAPS), reducing agents (DTT), carrier ampholytes and most of the time protease and phosphatase inhibitor cocktails are mandatory.

2.2. First Dimension: IEF with Immobilized Ph Gradients (Ipgs)

Proteins are amphoteric molecules; they carry positive, negative or zero net charge, depending on their amino acid composition. The net charge of a protein is the sum of all the negative and positive charges. The isoelectric point (pI) of a protein is the specific pH at which the net charge of the protein is zero. Proteins are positively charged at pH values below their pI and are negatively charged at pH values above their pI. IEF is an electrophoretic separation based on this specific biochemical characteristic of proteins.

Basically, the first dimension of the 2DE is achieved with a "strip". It is a dry gel that is formed by the polymerization of acrylamide monomers, linked by bis-acrylamide with molecules of covalently linked immobilin. Immobilins are chemical components that are derived from acrylamide and have additional ionizable non-amphoteric functions. Immobilins of various pKa can create an immobilized pH gradient inside the acrylamide gel. Immobilin was developed by Professors Righetti and Görg at the beginning of the 1990s and is now widely used in 2DE because the IEF gradient is very stable over time and in a high electric field, and shows good reproducibility and a large capacity for separation (9, 39-46).

The strip acrylamide gels are dried and cast on a plastic backing. Prior to use, they are rehydrated in a solution containing a pI-corresponding cocktail of carrier ampholytes and with the correct amount of proteins in the solubilization buffer. The carrier ampholytes are amphoteric molecules with a high buffering capacity near their pI. Commercial carrier ampholyte mixtures, which comprise species with pIs spanning a specific pH range, help the proteins to move.

When an electric field is applied, the negatively charged molecules (proteins and ampholytes) move towards the anode (positive / red electrode) and the positively charged molecules move towards the cathode (negative / black electrode). When the proteins are aligned according to their pI, the global net charge is zero and the protein is unable to move and is then focused. Focusing is achieved with a dedicated apparatus that is able to deliver up to 8000 or 10,000 V, but with a limitation in current intensity (50 μA maximum/strip) to reduce heat. The strips are usually first rehydrated without current for at least 5 h (passive rehydration), rehydrated with 50 V for 5 h (active rehydration) and then focused until at least 30 to 80 kV/h.

2.3. Strip Equilibration

The equilibration step is critical for 2DE. In this step, the strips are saturated with sodium dodecyl sulfate (SDS), an anionic detergent that can denature proteins and form a negatively charged protein / SDS complex. The amount of SDS bound to a protein is directly proportional to the mass of the protein. Thus, proteins that are completely covered by negative charges are separated on the basis of molecular mass.

The equilibration solution also contains buffer, with urea and glycerol. Equilibration of the strips is achieved in two steps: (1) with an equilibration solution containing DTT, to maintain a reducing environment; and (2) with an equilibration solution containing iodoacetamide, to alkylate reduced thiol groups, preventing their re-oxidation during electrophoresis.

2.4. Second Dimension: SDS-PAGE

In SDS polyacrylamide gel electrophoresis (SDS-PAGE), migration is determined not by the intrinsic electric charge of polypeptides but by their molecular weight. The SDS-denatured and reduced proteins are separated according to an apparent molecular weight, in comparison with a molecular weight marker. A linear relationship between the logarithm of the molecular weight and the distance of migration of the proteins can be used; it depends essentially on the percentage of polyacrylamide.

Equilibrated strips are embedded with 1% (w/v) low-melting-point agarose in TRIS / Glycine / SDS running buffer and with 0.01% bromophenol blue on the top of the second dimension acrylamide gel. Gels are usually run with 1 or 2 W of current in the first hour, followed by 15 mA/gel overnight with a temperature regulation (10°C to 18°C). When the bromophenol blue migration front reaches the bottom of the gel, the second dimension is finished and the acrylamide gel can be removed from the glass plates.

2.5. Protein Detection and Quantification

The gel must firstly be immersed in a fixation solution containing acid (phosphoric acid or acetic acid) and alcohol (ethanol or methanol) as a function of the staining protocol selected. Numerous stains can be used, but with very different costs (17). Conventional "visible" dyes are Coomassie Blue, colloidal Coomassie Blue and silver nitrate, with quite different sensitivities: 50, 10 and 0.5 ng of detectable protein/spot respectively (17, 20, 25, 47-51). Commercially available fluorescent dyes, such as Sypro Ruby, Flamingo and Deep Purple, have sensitivities of about 1 ng of detectable protein/spot (21, 25, 28-30, 52-55). Fluorescent dyes have the advantage of a 4 log dynamic linear range but the disadvantage of being more expensive. In comparison with fluorescent dyes, silver nitrate stain has a dynamic linear range of only 1.5 log, and is not recommended for a gel comparison study.

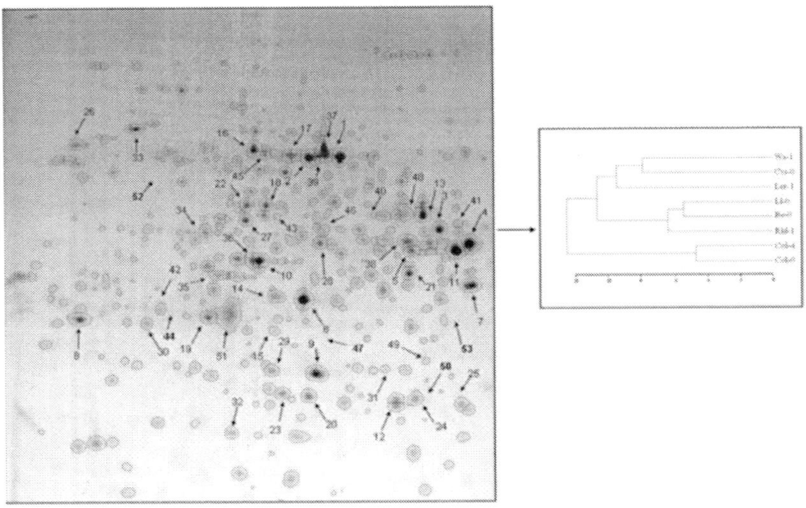

Figure 1. 2DE reference map of Arabidopsis thaliana soluble root proteins from ecotype Col-0 (left part). Proteins were separated using pH 4-7 IPG in first dimension and 11 % SDS-PAGE in second dimension. Proteins spots were visualized by colloidal Coomassie blue staining. The amount of each spot was estimated by its normalized volume as obtained by image analysis (65). Euclidian distances were then computed for all spots to build the similarity matrix for ecotypes, and clustering was performed using the Ward's method to link the variables (right part).

2.6. Computer-Assisted 2-D Image Analysis

Stained gels are scanned on a "visible" or "fluorescent" scanner as a function of the staining protocol selected. The image can then be imported to specific software to be analysed and compared. For a comparison study, at least three repetitions of the same sample should be run; many migration artifacts can occur during 2DE and, to reduce such variability, a mean of several gels is essential. Software, such as Image Master, Progenesis, PDQuest and Samespots, can be used to detect spots and to compare the spot intensity between samples (56-63). Spots of interest, *i.e.* spots specific to a sample or spots over-expressed on a condition/treatment, can be selected for further MS analysis. Several "computer-based" comparisons can be performed with a 2DE map. As a proteomic map is specific of a given cell, tissue or organism in a specific physiological condition, it is possible to compare not only one spot to one spot, but a set of spots to a set of spots, for example between two closed

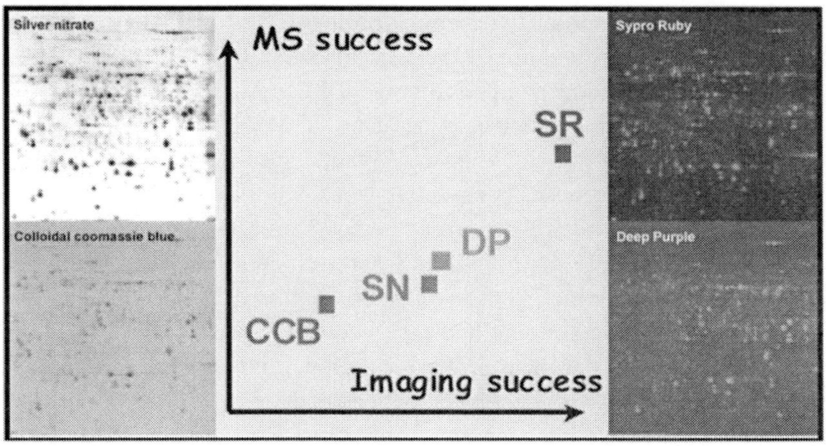

Figure 2. In these studies, the effects of protein spot properties were integrated to derive prediction of the MS results obtainable with the different dyes for all spots in the gels (24, 49). 100 μg of total protein extracts from Arabidopsis were focused on pI 4–7 range and separated on gels covering the 15–150 kDa range. By comparison to sensitivity properties of dyes, these simulations enable a first estimation of the overall proteomic capacity of dyes. They argue for a clear advantage in using fluorescent dyes, particularly SR, which cumulates high sensitivity, acceptable identification success on gels loaded with low protein amount, and constant protein sequence coverage. Abbreviations used: colloidal Coomassie blue (CCB), silver nitrate (SN), Sypro Ruby (SR), Deep Purple (DP).

organisms. In a precedent study, we investigated the natural variation in the proteome among 8 *Arabidopsis thaliana* ecotypes, of which 3 were previously shown to display atypical responses to environmental stress (64). The 2DE proteomic maps revealed important variations in terms of function between ecotypes (65).

Hierarchical clustering of proteomes according to either the amount of all anonymous spots, that of the 25 major spots or the functions of these major spots identified the same classes of ecotypes, and grouped the three atypical ecotypes (Fig. 1).

2.7. Protein Identification from 2-D Gel Spots

To identify the proteins within the spots of interest (according to image analysis), a gel with a greater amount of protein is prepared. In this case, IEF step must be performed at least until 100 kV/h. The other steps of the 2DE are very similar to the previously described protocol. Colloidal Coomassie Blue or fluorescent dyes are recommended for the staining of the preparative gel, because they have good compatibility with MS (22, 23, 28, 66). In contrast, silver nitrate will give poor results, even if MS-compatible protocols are available (21, 49, 50). It should be noted that a specific spot picker robot, able to work with fluorescence, is essential when working with fluorescent dyes. On a precedent study, we analyzed the total protein maps visualized when using classical visible stains and different fluorescent dyes (49). For this purpose, a soluble extract from *Arabidopsis thaliana* was taken as a model of sequenced eukaryotic genome and resolved by 2-DE. Besides specificities in background quality, propensity to saturation, and staining reproducibility, large differences were observed between dyes in terms of sensitivity, especially for low abundance spots. The effects of the staining procedure on MALDI-TOF MS characterization was analyzed too on a set of 48 protein spots that were selected for their contrasting abundance, pI, and Mr. Gels were stained with either classical visible stains colloidal (Coomassie blue and silver nitrate), and different fluorescent dyes (Sypro Ruby and Deep Purple). It appeared that Sypro Ruby combined several favorable features: no dependence of the identification rate upon the physicochemical properties of proteins, no impact on frequency of missed cleavages, and a higher predicted identification rate (Fig. 2).

3 - Recent Advances in "Gel-Based Proteomic" Approaches: Difference in Gel Electrophoresis (2D-DIGE)

Difference in gel electrophoresis (DIGE), first conceived by Unlu et al. in 1997, takes advantages of structurally similar cyanine-based dyes to label different pools of protein samples, which are then co-separated on a single 2DE gel (34).

The biggest advantage of DIGE over other two-dimensional- based technologies is that it enables the analysis of two or more protein samples simultaneously on a single 2DE (31, 32, 35, 36). Because the same proteins present in two different samples were prelabeled with two different dyes (i.e., Cy3 and Cy5, respectively), they could be combined and separated on the same 2DE without the loss of the relative protein abundance in the original samples. At the end of protein separation, the relative ratio of proteins in the two original samples could be readily obtained by comparing the fluorescence intensity of the same protein spots under different detection channels (e.g., Cy3 and Cy5) using a commercial fluorescence gel scanner. Because only one gel is used in DIGE, and the same proteins from two different protein samples comigrate as single spots, there is no need for the generation of "averaged" gels, as well as superimposition of different gels, making spot comparison and protein quantitation much more convenient and reliable. This makes DIGE potentially amendable for high-throughput proteomics applications.

DIGE has shown significant advantages over conventional 2DE in a number of applications. Up to three kinds of fluorescent cyanine dyes have been employed in DIGE, namely, Cy2, Cy3, and Cy5, which allows for simultaneous analysis of up to three different protein samples in a single gel. DIGE is a valuable method for high-throughput studies of protein expression profiles, providing opportunities to detect and quantify accurately "difficult" proteins, such as low-abundance proteins.

A problem in DIGE lies in the hydrophobicity of the cyanine dyes, which label the protein by reacting covalently, to a large extent, with surface-exposed lysines in the protein, and lead to removal of multiple charges from the protein. Consequently, this decreases the solubility of the labeled protein, and in some cases may lead to protein precipitation prior to gel electrophoresis. To address this problem, minimal labeling is generally employed in DIGE. Typically the labelling reaction is optimized such that only 1–5% of total lysines in a given protein are labeled. Alternatively, Shaw et al.

have developed a new batch of DIGE Cy3 and Cy5 dyes, which label only free cysteines in a protein by "saturation" labeling (37). This strategy offers greater sensitivity than the conventional DIGE method. The biggest drawback, however, is that it only labels proteins that contain free cysteines, meaning that a certain percentage of proteins in a proteome will not be labeled with this strategy, let alone downstream detection and characterization of these proteins.

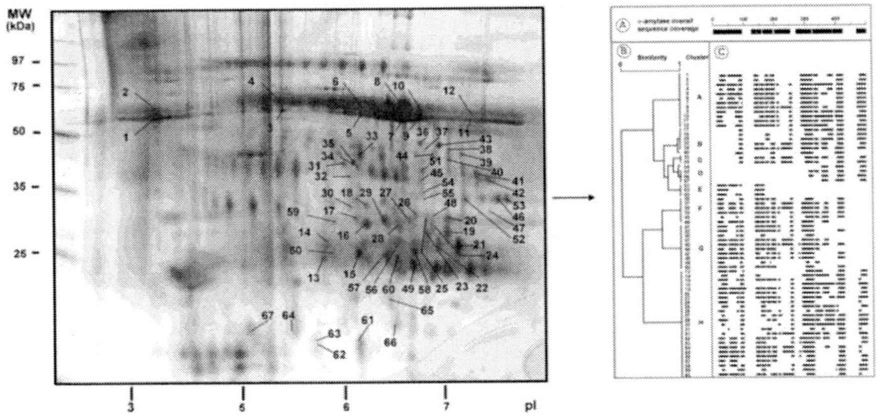

Figure 3. 2DE map of human salivary proteins (71). Proteins were resolved using pH 3–10NL IPG and 12% SDS-PAGE, and stained with colloidal Coomassie blue. Alpha-amylase spots subsequently identified by MALDI-TOF MS are indicated by respective spot number (left part). According to the mass features of alpha-amylase (right part) identified spots, (A) coverage of the alpha-amylase sequence by the total population of peptides identified (black boxes) in the different alpha-amylase spots, (B) simultaneous clustering of the 67 alpha-amylase spots according to the MW range measured on gels, the MS identification of peptides in the N-terminal and C-terminal and central regions of the sequence, (C) individual spot coverage of the alpha-amylase sequence by peptides identified (black boxes) by MALDI-TOF MS.

4 - Benefits of "Gel-Based Proteomic" Approaches: Characterization of Protein Isoforms

An area of increasing interest in proteomics is the identification of post-translational modifications and / or spliced forms of a same gene or protein (67-70). The process of determining whether a protein is expressed in a particular proteome has become a relatively simple task with the automation of the 'in-gel' digest and subsequent identification of the resulting peptides with mass spectrometers. Today, most proteins are identified by either assigning them definitive protein attributes, such as peptide masses generated by MALDI–TOF mass spectrometry and the short amino acid sequences generated by tandem MS. It is clear that when several spliced variants are present in a proteome, such approach for protein identification mostly characterizes peptides common to all spliced variants. In a precedent study, we used the advantages of 2DE separation to analyze alpha-amylase diversity in human saliva (71). Because each alpha-amylase isoforms exist as a discrete purified protein, any information obtained from the analysis of this protein is unique to its original proteome (Fig. 3). 2DE was combined with systematic MALDI-TOF MS analysis and more than 140 protein spots identifying the alpha-amylase were shown to constitute a stable but very complex pattern.

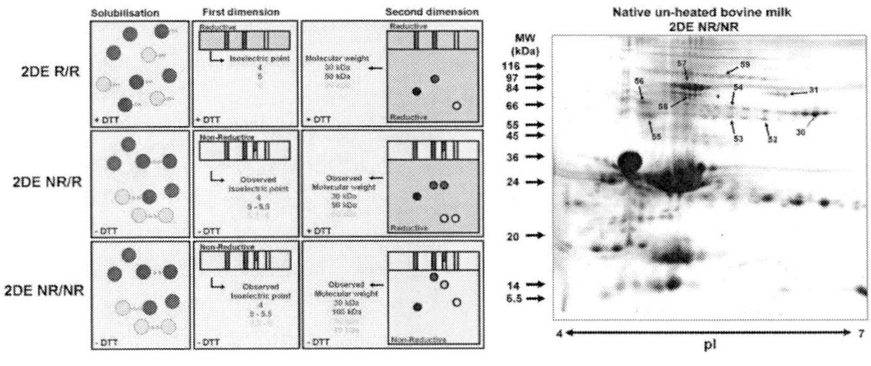

Figure 4. Three 2DE conditions were set up to analyze and compare disulphide bridge exchanges between milk proteins (left part). 2DE R/R: samples completely reduced before and during 2-dimensional electrophoresis. 2DE NR/R: samples un-reduced before 2-dimensional electrophoresis and reduced only after isoelectric focusing. 2DE NR/NR: samples unreduced before and during 2-dimensional electrophoresis. The corresponding 2DE of proteins in raw milk (100 µg) separated under non-reducing conditions (NR/NR) using a 7-cm pH 4-7 pI range strip for the first dimension, and a 10 to 18% gradient acrylamide gel for the second dimension (Right part). The specific/interesting spots as indicated by arrows were submitted to MALDI-TOF mass to identify proteins involved in polymers (72).

Careful analysis of mass spectra and simultaneous hierarchical clustering of the observed peptides and of the electrophoretic features of spots defined several groups of isoforms (Fig. 3 right part) with specific sequence characteristics, potentially related with special biological activities. In a recent study, 2DE separation was successfully used to analyze isoforms and polymers forms of bovine milk proteins (72).

A combination of reducing and non-reducing steps was used to reveal proteins polymers occurring before or after heat treatment of milk (Fig. 4). This original 2DE strategy revealed numerous disulfide-mediated interactions and was proposed to analyze reduction/oxidation of milk and dairy product proteins.

5 - Limits and Answers of "Gel-Based Proteomic" Approaches

5.1 - Low-Abundance Proteins

Low-abundance proteins are rarely seen on traditional 2D maps because large quantities of abundant soluble proteins obscure their detection (21, 73-75). Most 2DE-based proteomic studies employ a 'one-extract–one-gel' approach and the majority of proteins identified in these studies are in high abundance. These low-abundance proteins are considered to be some of the most important, including regulatory proteins, signal transduction proteins and receptors. Consequently, the analysis of low-abundance proteins is becoming increasingly common in cellular proteomics. The dynamic range of protein concentration can differ considerably between biological samples. For yeast, the most abundant proteins are present at around 2 000 000 copies per cell, whereas the least abundant proteins are present at around 100 copies per cell, a dynamic range of only 4 orders of magnitude. However, in plasma, the predicted dynamic range of proteins is up to 12 orders of magnitude. Analysis of individual compartments not only provides information on protein localization, but also allows detection of protein populations otherwise not detectable in whole cell proteomes. Detection of the low-abundance proteins requires most of the time removal of abundant proteins from the sample. For example, the complexity of the serum and plasma proteome presents extreme analytical challenges in comprehensive analysis due to the wide dynamic range of protein concentrations. Therefore, robust sample preparation methods remain one of the important steps in the proteome characterization workflow. A specific depletion of high-abundant proteins from human serum and plasma using affinity columns is of particular interest to improve dynamic range for proteomic analysis and enable the identification of low-abundant plasma proteins (74, 76). On another hand, IPG technology can be used with narrow (2–3 pH units) and very narrow (~1 pH unit) gradients that enable many more proteins to be resolved. Indeed, the advent of immobilized pH gradients has greatly improved the reproducibility of 2D gels and has made it easier for new users to implement this technology. The loading capacity of narrow-range IPGs is considerably higher than broad-range IPGs, thus enabling the visualization and identification of previously unseen proteins. Sub-fractionation fractionation can be used to improve the recovery of low-abundance proteins too. For example, membrane preparation methods are commercially available and allow a specific separation between abundant / soluble proteins and membrane / low-abundance proteins. More recently, a system is available to perform a specific depletion of high-abundant proteins and a reduction of protein concentration differences (77, 78). The protein population is "equalized", by sharply reducing the concentration of the most abundant components, while simultaneously enhancing the concentration of the most dilute species.

5.2 - Membrane Proteins

The resolution of membrane proteins remains an area of considerable concern in gel-based proteomics (79-84). There remains an attitude that it is difficult or impossible to effectively resolve membrane proteins using 2DE. Indeed, few membrane proteins are seen on 2D gels when conventional sample-preparation methods are used. Membrane proteins are poorly soluble in the detergent / chaotrope conditions available for IEF, and are inherently insoluble in gel matrices under these conditions and thus are poorly resolved by IEF and subsequent 2DE. Fractionation, in combination with the correct solubilizing reagents, produces sample extracts that are highly enriched for membrane proteins. Sequential extraction of proteins from a sample by increasing protein solubility at each step is an effective strategy for first removing the more abundant soluble proteins and then for concentrating the less abundant and less soluble membrane proteins.

5.3 – Extreme pI Proteins: The Case of Alkaline Proteins

Alkaline proteins were particularly difficult to resolve on 2D gels. First, the most common commercially available pH gradients, until recently, were pH 4–7 and pH 3–10 and these do not provide significant alkaline-protein resolution. As more alkaline pH range immobilized pH gradients become commercially available, resolution of proteins in IPGs up to pH 12 has been demonstrated. Strongly alkaline proteins such as ribosomal and nuclear proteins with closely related pIs between 10.5 and 11.8 were focused to the steady state by using 3-12, 6-12 and 9– 12 pI ranges (85-87). For highly resolved 2-D patterns, different optimization steps with respect to pH engineering and gel composition were necessary, such as the substitution of dimethylacrylamide for acrylamide, the addition near the cathode of a paper strip soaked with DTT providing a continuous influx of DTT to compensate for the loss of DTT (41, 45), and the addition of isopropanol to the IPG rehydration solution in order to suppress the reverse electroendosmotic flow which causes highly streaky 2-D patterns in narrow pH range IPGs 9-12 and 10-12 (87).

6 – What Future for "Gel-Based Proteomic" Approaches?

Thanks to its high resolving power and its large sample loading capacity, 2DE allows several hundred proteins to be displayed simultaneously on a single gel, producing a direct and global view of a sample proteome at a given time point. Reference maps of numerous distinct samples have now been published, providing, to researchers worldwide, standardized libraries of proteins known to be present in these samples. But 2DE has some limitations that must be taken into account. Despite maximal precautions, there will be some degree of gel-to-gel and run-to-run variability in the expression of the same protein set, which could be overcome by maintaining a variability coefficient of reference spots as low as possible. It can be largely circumvented using a DIGE strategy. Additionally, some proteins may escape the capabilities of conventional 2DE for several reasons, including the poor solubility of membrane proteins and out of range characteristics of extreme proteins such as high or low pI and molecular weight. Despite all these drawbacks, 2DE can demonstrate changes in relative abundance of visualized proteins and can detect protein isoforms, variants, polymer complexes and posttranslational modifications. Quantitative proteomics can be achieved by assessing differences in protein expression across gels using 2DE dedicated software and proteins in varying spots can be identified by MS. The uniqueness of 2DE for easy visualization of protein isoforms renders this technology itself extremely informative and it is currently the most rapid method for direct targeting of protein expression differences.

REFERENCES

[1] O'Farrell PH. High resolution two-dimensional electrophoresis of proteins. *Journal of Biological Chemistry.* 1975 May 25, 1975; 250(10):4007-21.

[2] Wan JH, He FC. Technical development of proteomics. *Chinese Science Bulletin.* 1999 Aug;44(16):1441-7.

[3] Gromov PS, Celis JE. From genomics to proteomics. *Molecular Biology.* 2000 Jul-Aug;34(4):508-20.

[4] Govorun VM, Archakov AI. Proteomic technologies in modern biomedical science. *Biochemistry-Moscow.* 2002 Oct;67(10):1109-23.

[5] Garfin DE. Two-dimensional gel electrophoresis: an overview. *Trac-Trends in Analytical Chemistry.* 2003 May;22(5):263-72.

[6] Collinsova M, Jiracek J. Current development in proteomics. *Chemicke Listy.* 2004;98(12):1112-8.

[7] Bradshaw RA, Burlingame AL. From proteins to proteomics. *Iubmb Life.* [Review]. 2005 Apr-May;57(4-5):267-72.

[8] Van den Bergh G, Arckens L. Recent advances in 2D electrophoresis: an array of possibilities. *Expert Review of Proteomics.* 2005 Apr;2(2):243-52.

[9] Carrette O, Burkhard PR, Sanchez JC, Hochstrasser DF. State-of-the-art two-dimensional gel electrophoresis: a key tool of proteomics research. *Nature Protocols.* [Article]. 2006;1(2):812-23.

[10] Bergeron JJM, Bradshaw RA. What has proteomics accomplished? *Molecular & Cellular Proteomics.* [Editorial Material]. 2007 Oct; 6(10):1824-6.

[11] Penque D. Two-dimensional gel electrophoresis and mass spectrometry for biomarker discovery. *Proteomics Clinical Applications.* 2009 Feb;3(2):155-72.

[12] Van den Bergh G, Arckens L. High Resolution Protein Display by Two-Dimensional Electrophoresis. *Current Analytical Chemistry*. 2009 Apr;5(2):106-15.

[13] HumpherySmith I, Cordwell SJ, Blackstock WP. Proteome research: Complementarity and limitations with respect to the RNA and DNA worlds. *Electrophoresis*. 1997 Aug;18(8):1217-42.

[14] Celis JE, Gromov P. 2D protein electrophoresis: can it be perfected? *Current Opinion in Biotechnology*. 1999 Feb;10(1):16-21.

[15] Ong SE, Pandey A. An evaluation of the use of two-dimensional gel electrophoresis in proteomics. *Biomolecular Engineering*. 2001 Nov;18(5):195-205.

[16] Lopez JL. Two-dimensional electrophoresis in proteome expression analysis. *Journal of Chromatography B-Analytical Technologies in the Biomedical and Life Sciences*. 2007 Apr;849(1-2):190-202.

[17] Chevalier F, Rofidal V, Rossignol M. *Visible and fluorescent staining of two-dimensional gels*2007.

[18] Harris LR, Churchward MA, Butt RH, Coorssen JR. Assessing detection methods for gel-based proteomic analyses. *Journal of Proteome Research*. 2007;6(4):1418-25.

[19] Volkova KD, Kovalska VB, Yarmoluk SM. Modern techniques for protein detection on polyacrylamide gels: problems arising from the use of dyes of undisclosed structures for scientific purposes. *Biotechnic & Histochemistry*. 2007 Aug-Oct;82(4-5):201-8.

[20] Smejkal GB. The Coomassie chronicles: past, present and future perspectives in polyacrylamide gel staining. *Expert Review of Proteomics*. 2004 Dec;1(4):381-7.

[21] Patton WF. Detection technologies in proteome analysis. *Journal of Chromatography B-Analytical Technologies in the Biomedical and Life Sciences*. [Review]. 2002 May;771(1-2):3-31.

[22] Cong WT, Hwang SY, Jin LT, Choi JK. Sensitive fluorescent staining for proteomic analysis of proteins in 1-D and 2-D SDS-PAGE and its comparison with SYPRO Ruby by PMF. *Electrophoresis*. 2008 Nov;29(21):4304-15.

[23] Ball MS, Karuso P. Mass spectral compatibility of four proteomics stains. *Journal of Proteome Research*. 2007 Nov;6(11):4313-20.

[24] Chevalier F, Centeno D, Rofidal V, Tauzin M, Martin O, Sommerer N, et al. Different impact of staining procedures using visible stains and fluorescent dyes for large-scale investigation of proteomes by MALDI-

TOF mass spectrometry. *Journal of Proteome Research*. 2006 Mar;5(3):512-20.

[25] Lanne B, Panfilov O. Protein staining influences the quality of mass spectra obtained by peptide mass fingerprinting after separation on 2-D gels. A comparison of staining with coomassie brilliant blue and SYPRO Ruby. *Journal of Proteome Research*. 2005 Jan-Feb;4(1):175-9.

[26] Lamanda A, Zahn A, Roder D, Langen H. Improved Ruthenium II tris (bathophenantroline disulfonate) staining and destaining protocol for a better signal-to-background ratio and improved baseline resolution. *Proteomics*. 2004 Mar;4(3):599-608.

[27] White IR, Pickford R, Wood J, Skehel JM, Gangadharan B, Cutler P. A statistical comparison of silver and SYPRO Ruby staining for proteomic analysis. *Electrophoresis*. 2004 Sep;25(17):3048-54.

[28] Berggren KN, Schulenberg B, Lopez MF, Steinberg TH, Bogdanova A, Smejkal G, et al. An improved formulation of SYPRO Ruby protein gel stain: Comparison with the original formulation and with a ruthenium II tris (bathophenanthroline disulfonate) formulation. *Proteomics*. [Article]. 2002 May;2(5):486-98.

[29] Rabilloud T, Strub JM, Luche S, van Dorsselaer A, Lunardi J. Comparison between Sypro Ruby and ruthenium II tris (bathophenanthroline disulfonate) as fluorescent stains for protein detection in gels. *Proteomics*. [Article]. 2001 May;1(5):699-704.

[30] Berggren K, Steinberg TH, Lauber WM, Carroll JA, Lopez MF, Chernokalskaya E, et al. A luminescent ruthenium complex for ultrasensitive detection of proteins immobilized on membrane supports. *Analytical Biochemistry*. [Article]. 1999 Dec;276(2):129-43.

[31] Karp NA, Feret R, Rubtsov DV, Lilley KS. Comparison of DIGE and post-stained gel electrophoresis with both traditional and SameSpots analysis for quantitative proteomics. *Proteomics*. 2008 Mar;8(5):948-60.

[32] Hrebicek T, Duerrschmid K, Auer N, Bayer K, Rizzi A. Effect of CyDye minimum labeling in differential gel electrophoresis on the reliability of protein identification. *Electrophoresis*. 2007 Apr;28(7):1161-9.

[33] Karp NA, McCormick PS, Russell MR, Lilley KS. Experimental and statistical considerations to avoid false conclusions in proteomics studies using differential in-gel electrophoresis. *Molecular & Cellular Proteomics*. 2007 Aug;6(8):1354-64.

[34] Viswanathan S, Unlu M, Minden JS. Two-dimensional difference gel electrophoresis. *Nature Protocols*. 2006;1(3):1351-8.

[35] Wheelock AM, Morin D, Bartosiewicz M, Buckpitt AR. Use of a fluorescent internal protein standard to achieve quantitative two-dimensional gel electrophoresis. *Proteomics*. 2006 Mar;6(5):1385-98.

[36] Gade D, Thiermann J, Markowsky D, Rabus R. Evaluation of two-dimensional difference gel electrophoresis for protein profiling. *Journal of Molecular Microbiology and Biotechnology*. [Article]. 2003;5(4):240-51.

[37] Shaw J, Rowlinson R, Nickson J, Stone T, Sweet A, Williams K, et al. Evaluation of saturation labelling two-dimensional difference gel electrophoresis fluorescent dyes. *Proteomics*. [Article]. 2003 Jul;3(7):1181-95.

[38] Westermeier R, Loyland S, Asbury R. Proteomics technology. Journal of Clinical Ligand Assay. 2002 Fal;25(3):242-52.

[39] Bjellqvist B, Ek K, Righetti PG, Gianazza E, Gorg A, Westermeier R, et al. Isoelectric-focusing in immobilized ph gradients - principle, methodology and some applications. *Journal of Biochemical and Biophysical Methods*. [Article]. 1982;6(4):317-39.

[40] Righetti PG, Castagna A, Hamdan M. Recent trends in proteome analysis. *Advances in Chromatography*, Vol 422003. p. 269-321.

[41] Altland K, Becher P, Rossmann U, Bjellqvist B. Isoelectric-focusing of basic-proteins - the problem of oxidation of cysteines. *Electrophoresis*. [Article]. 1988 Sep;9(9):474-85.

[42] Bouchal P, Kucera I. Two-dimensional electrophoresis in proteomics: Principles and applications. Chemicke Listy. 2002;97(1):29-36.

[43] Friedman DB, Hoving S, Westermeier R. Isoelectric focusing and two-dimensional gel electrophoresis. *Guide to Protein Purification,* Second Edition2009. p. 515-40.

[44] Gorg A, Postel W, Gunther S. The current state of two-dimensional electrophoresis with immobilized ph gradients. *Electrophoresis*. [Review]. 1988 Sep;9(9):531-46.

[45] Gorg A, Boguth G, Obermaier C, Posch A, Weiss W. 2-dimensional polyacrylamide-gel electrophoresis with immobilized ph gradients in the first dimension (ipg-dalt) - the state-of-the-art and the controversy of vertical versus horizontal systems. *Electrophoresis*. [Proceedings Paper]. 1995 Jul;16(7):1079-86.

[46] Gorg A, Weiss W, Dunn MJ. Current two-dimensional electrophoresis technology for proteomics. *Proteomics*. [Review]. 2004 Dec;4(12):3665-85.

[47] Rabilloud T. A comparison between low background silver diammine and silver-nitrate protein stains. *Electrophoresis*. [Article]. 1992 Jul;13(7):429-39.

[48] Neuhoff V, Stamm R, Pardowitz I, Arold N, Ehrhardt W, Taube D. Essential problems in quantification of proteins following colloidal staining with coomassie brilliant blue dyes in polyacrylamide gels, and their solution. *Electrophoresis*. [Article]. 1990 Feb;11(2):101-17.

[49] Chevalier F, Rofidal V, Vanova P, Bergoin A, Rossignol M. Proteomic capacity of recent fluorescent dyes for protein staining. *Phytochemistry*. 2004 Jun;65(11):1499-506.

[50] Mortz E, Krogh TN, Vorum H, Gorg A. Improved silver staining protocols for high sensitivity protein identification using matrix-assisted laser desorption/ionization-time of flight analysis. *Proteomics*. [Article]. 2001 Nov;1(11):1359-63.

[51] Jin LT, Li XK, Cong WT, Hwang SY, Choi JK. Previsible silver staining of protein in electrophoresis gels with mass spectrometry compatibility. *Analytical Biochemistry*. 2008 Dec;383(2):137-43.

[52] Bell PJL, Karuso P. Epicocconone, a novel fluorescent compound from the fungus Epicoccum nigrum. *Journal of the American Chemical Society*. [Article]. 2003 Aug;125(31):9304-5.

[53] Mackintosh JA, Choi HY, Bae SH, Veal DA, Bell PJ, Ferrari BC, et al. A fluorescent natural product for ultra sensitive detection of proteins in one-dimensional and two-dimensional gel electrophoresis. *Proteomics*. [Proceedings Paper]. 2003 Dec;3(12):2273-88.

[54] Ahnert N, Patton WF, Schulenberg B. Optimized conditions for diluting and reusing a fluorescent protein gel stain. *Electrophoresis*. 2004 Aug;25(15):2506-10.

[55] Westermeier R, Marouga R. Protein detection methods in proteomics research. *Bioscience Reports*. 2005 Feb;25(1-2):19-32.

[56] Rosengren AT, Salmi JM, Aittokallio T, Westerholm J, Lahesmaa R, Nyman TA, et al. Comparison of PDQuest and Progenesis software packages in the analysis of two-dimensional electrophoresis gels. *Proteomics*. 2003 Oct;3(10):1936-46.

[57] Nebrich G, Liegmann H, Wacker M, Herrmann M, Sagi D, Landowsky A, et al. Proteomer, a novel software application for management of proteomic 2DE-gel data-II application. *Molecular & Cellular Proteomics*. 2005 Aug;4(8):S296-S.

[58] Wheelock AM, Buckpitt AR. Software-induced variance in two-dimensional gel electrophoresis image analysis. *Electrophoresis*. 2005 Dec;26(23):4508-20.

[59] Maurer MH. Software analysis of two-dimensional electrophoretic gels in proteomic experiments. *Current Bioinformatics*. 2006 May;1(2):255-62.

[60] Wheelock AM, Goto S. Effects of post-electrophoretic analysis on variance in gel-based proteomics. *Expert Review of Proteomics*. 2006 Feb;3(1):129-42.

[61] Clark BN, Gutstein HB. The myth of automated, high-throughput two-dimensional gel analysis. *Proteomics*. 2008 Mar;8(6):1197-203.

[62] Panchaud A, Affolter M, Moreillon P, Kussmann M. Experimental and computational approaches to quantitative proteomics: Status quo and outlook. *Journal of Proteomics*. 2008 Apr;71(1):19-33.

[63] Kang YY, Techanukul T, Mantalaris A, Nagy JM. Comparison of Three Commercially Available DIGE Analysis Software Packages: Minimal User Intervention in Gel-Based Proteomics. *Journal of Proteome Research*. 2009 Feb;8(2):1077-84.

[64] Chevalier F, Pata M, Nacry P, Doumas P, Rossignol M. Effects of phosphate availability on the root system architecture: large-scale analysis of the natural variation between Arabidopsis accessions. *Plant Cell and Environment*. 2003 Nov;26(11):1839-50.

[65] Chevalier F, Martin O, Rofidal V, Devauchelle AD, Barteau S, Sommerer N, et al. Proteomic investigation of natural variation between Arabidopsis ecotypes. *Proteomics*. 2004 May;4(5):1372-81.

[66] Nock CM, Ball MS, White IR, Skehel JM, Bill L, Karuso P. Mass spectrometric compatibility of Deep Purple and SYPRO Ruby total protein stains for high-throughput proteomics using large-format two-dimensional gel electrophoresis. *Rapid Communications in Mass Spectrometry*. 2008;22(6):881-6.

[67] Holland JW, Deeth HC, Alewood PF. Proteomic analysis of K-casein micro-heterogeneity. *Proteomics*. 2004 Mar;4(3):743-52.

[68] Schulenberg B, Goodman TN, Aggeler R, Capaldi RA, Patton WF. Characterization of dynamic and steady-state protein phosphorylation using a fluorescent phosphoprotein gel stain and mass spectrometry. *Electrophoresis*. 2004 Aug;25(15):2526-32.

[69] Ahrer K, Jungbauer A. Chromatographic and electrophoretic characterization of protein variants. *Journal of Chromatography B-*

Analytical Technologies in the Biomedical and Life Sciences. 2006 Sep;841(1-2):110-22.

[70] Poth AG, Deeth HC, Alewood PF, Holland JW. Analysis of the Human Casein Phosphoproteome by 2-D Electrophoresis and MALDI-TOF/TOF MS Reveals New Phosphoforms. *Journal of Proteome Research*. 2008 Nov;7(11):5017-27.

[71] Hirtz C, Chevalier F, Centeno D, Rofidal V, Egea JC, Rossignol M, et al. MS characterization of multiple forms of alpha-amylase in human saliva. *Proteomics*. 2005 Nov;5(17):4597-607.

[72] Chevalier F, Hirtz C, Sommerer N, Kelly AL. Use of Reducing/ Nonreducing Two-Dimensional Electrophoresis for the Study of Disulfide-Mediated Interactions between Proteins in Raw and Heated Bovine Milk. *Journal of Agricultural and Food Chemistry*. 2009 Jul;57(13):5948-55.

[73] Yamada M, Murakami K, Wallingford JC, Yuki Y. Identification of low-abundance proteins of bovine colostral and mature milk using two-dimensional electrophoresis followed by microsequencing and mass spectrometry. *Electrophoresis*. [Article]. 2002 Apr;23(7-8):1153-60.

[74] Greenough C, Jenkins RE, Kitteringham NR, Pirmohamed M, Park BK, Pennington SR. A method for the rapid depletion of albumin and immunoglobulin from human plasma. *Proteomics*. [Article]. 2004 Oct;4(10):3107-11.

[75] Ahmed N, Rice GE. Strategies for revealing lower abundance proteins in two-dimensional protein maps. *Journal of Chromatography B-Analytical Technologies in the Biomedical and Life Sciences*. [Review]. 2005 Feb;815(1-2):39-50.

[76] Issaq HJ, Xiao Z, Veenstra TD. Serum and plasma proteomics. *Chemical Reviews*. 2007 Aug;107(8):3601-20.

[77] Righetti PG, Castagna A, Boschetti E, Lomas L. Equalizer beads; The quest for a democratic proteome. Molecular & Cellular *Proteomics*. 2005 Aug;4(8):S12-S.

[78] Righetti PG, Boschetti E, Lomas L, Citterio A. Protein Equalizer (TM) Technology: The quest for a democratic proteome. *Proteomics*. 2006 Jul;6(14):3980-92.

[79] Luche S, Santoni V, Rabilloud T. Evaluation of nonionic and zwitterionic detergents as membrane protein solubilizers in two-dimensional electrophoresis. *Proteomics*. 2003 Mar;3(3):249-53.

[80] Santoni V, Kieffer S, Desclaux D, Masson F, Rabilloud T. Membrane proteomics: Use of additive main effects with multiplicative interaction

model to classify plasma membrane proteins according to their solubility and electrophoretic properties. *Electrophoresis*. 2000 Oct;21(16):3329-44.

[81] Santoni V, Molloy M, Rabilloud T. Membrane proteins and proteomics: Un amour impossible? *Electrophoresis*. 2000 Apr;21(6):1054-70.

[82] Santoni V, Doumas P, Rouquie D, Mansion M, Rabilloud T, Rossignol M. Large scale characterization of plant plasma membrane proteins. *Biochimie*. 1999 Jun;81(6):655-61.

[83] Santoni V, Rabilloud T, Doumas P, Rouquie D, Mansion M, Kieffer S, et al. Towards the recovery of hydrophobic proteins on two-dimensional electrophoresis gels. *Electrophoresis*. 1999 Apr-May;20(4-5):705-11.

[84] Chevallet M, Santoni V, Poinas A, Rouquie D, Fuchs A, Kieffer S, et al. New zwitterionic detergents improve the analysis of membrane proteins by two-dimensional electrophoresis. *Electrophoresis*. 1998 Aug; 19(11):1901-9.

[85] Drews O, Reil G, Parlar H, Gorg A. Setting up standards and a reference map for the alkaline proteome of the Gram-positive bacterium Lactococcus lactis. *Proteomics*. [Proceedings Paper]. 2004 May;4 (5):1293-304.

[86] Wildgruber R, Reil G, Drews O, Parlar H, Gorg A. Web-based two-dimensional database of Saccharomyces cerevisiae proteins using immobilized pH gradients from pH 6 to pH 12 and matrix-assisted laser desorption/ionization-time of flight mass spectrometry. *Proteomics*. [Proceedings Paper]. 2002 Jun;2(6):727-32.

[87] Gorg A, Obermaier C, Boguth G, Csordas A, Diaz JJ, Madjar JJ. Very alkaline immobilized pH gradients for two-dimensional electrophoresis of ribosomal and nuclear proteins. *Electrophoresis*. [Proceedings Paper]. 1997 Mar-Apr;18(3-4):328-37.

In: Proteomics
Editor: Giselle C. Rancourt

ISBN: 978-1-61668-691-8

Chapter 4

Algorithms for the Quantification of Proteins from High-throughput Liquid Chromatography-mass Spectrometry (LC-MS) Data

Ole Schulz-Trieglaff
Free University of Berlin, Institute for Computer Science,
Berlin, Germany

Abstract

In this commentary, we review algorithms for the analysis of liquid chromatography-mass spectrometry (LC-MS) data. Mass spectrometry is a technology that can be used to determine the identities and abundances of the compounds in complex samples. In combination with liquid chromatography, it has become a popular method in the field of proteomics, the large-scale study of proteins and peptides in living systems.

The data sets obtained from an LC-MS experiment are large and highly complex. The outcome of such an experiment is called an LC-MS map and is a collection of mass spectra. They contain, among the signals of interest, a high amount of noise and other disturbances. That is why algorithms for the low-level processing of LC-MS data are becoming increasingly important.

In this commentary, we revied the state-of-the-art of quantification algorithms, their capabilities and also limitations and outline avenues for future research.

1. Introduction

Proteins are key players in the cell: they are involved in signaling pathways, cell death and crucial body function such as muscle contraction. In addition, proteins circulate through the body and interact with most organs. Therefore, proteins are also thought to be excellent biomarker for clinical studies. A deeper knowledge of their function is crucial for the development of new drugs and thus the cure of diseases. Accordingly, proteomics as a scientific discipline has received a lot of attention in recent years but also meets growing interest in the pharmaceutical industry. In sum, this field aims at a comprehensive understanding of the protein content of a biological specimen, the study of sequence, structure and abundance of all proteins at a given time and condition.

A typical proteomics experiment requires the analysis of highly complex biological data, i.e., a sample containing many proteins at varying concentrations. In most cases, proteins will only be one class of molecules among many other sample compounds. Established technologies, such as electrophoresis-based methods, have problems to cope with this complexity.

For these reasons, mass spectrometry-based methods have become increasingly popular. This technology can quantify and sequence proteins even in complex samples and in a high-throughput manner. This commentary deals with data from a specific type of mass spectrometry experiment in which a chromatographic column is coupled to the mass spectrometer. This setup is called a liquid chromatography-mass spectrometry (LC-MS) experiment.

Quality and quantity of the data generated from mass spectrometry experiments have increased dramatically in recent years. New instruments allow highly accurate mass measurements, but also generate vast amounts of data. These problems have lead the scientific community to the conclusion that efficient algorithms are crucial for progress in the field of mass spectrometry-based proteomics.

The next section gives a short introduction into LC-MS instrumention and explains the key principles of a mass spectrometer. These are crucial if we want to understand the full scope but also the limitations of MS-based proteomics. The third section gives an overview of algorithms for the analysis of LC-MS experiments. While this commentary focusses on quantitative proteomics, we will also briefly highlight algorithms for the indentification of peptides and proteins in a sample since they are usually applied in conjunction with quantification methods. We conclude by summarizing the state of the start of the field and by outlining future directions.

2. Instrumentation

The foundations of mass spectrometry (MS) have been developed as early as 1899 by the Germany physicist Wilhelm Wien. He devised an instrument that could separate positively charged ions by their charge-to-mass ratio (z/m) using electric fields [1]. This method was refined by the English scientist Sir Joseph John Thompson in the early 20th century. He also discovered that most chemical elements exist in different isotope states using a variant of the instrument devised by Wien.

Compared to those early beginnings, the application of mass spectrometry to complex proteomic samples and in a high-throughput fashion have just recently become popular in the scientific community. To give an example, in 2002, John Fenn and Koichi Tanaka received the Nobel Prize in Chemistry for the development of Electrospray Ionization (ESI) and of Soft Laser Desorption (SLD), respectively. SLD was later improved by the German scientists Franz Hillenkamp and Michael Karras and is now more widely known as matrix-assisted laser desorption/ionization (MALDI).

A typical mass spectrometry experiment in a modern laboratory involves the analysis of mixtures i.e. samples containing many proteins at different concentrations. We have to keep in mind that the majority of the low-abundance proteins are not observed in these experiments. High-throughput mass spectrometry experiments capture only a subset of all proteins contained in the sample. Furthermore, there is even some stochastic moment to these experiments as reproducibility between technical replicates can be low and not the same set of proteins will be detected every time.

The outcome of an LC-MS experiment is an *LC-MS map*, a set of mass spectra. Each spectrum is a set of tuples $p_i = (m/z, it)$. m/z is the mass-to-charge ratio and has the unit thomson (Th). it is the intensity. Its exact meaning depends on the instrument used, but in most cases we assume it to be an *ion count*. We use it as an estimate of relative abundance of the compound at the corresponding m/z. Note that the signal intensities of the detected peptides are only comparable between mass spectrometers of the same type and if care is taken during sample preparation.

2.1. Sample Preparation

The sample consisting of proteins and other substances is subjected to enzymatic digestion. This facilitates the later analysis as MS of whole proteins is less sensitive than peptide MS [2]. This is also called the *bottom-up approach*. In most cases, the enzyme *trypsin* is used for digestion since it cuts the amino acid sequence of a protein at well defined positions. So-called *miscleavages* might also occur, leading to peptides than span one or two tryptic cleavage sites.

Other enzymes are used to remove biomaterials such as lipids or sugars. Another step in a typical sample preparation is to break protein-protein interactions using urea or to reduce disulfid bonds between cysteine residues.

Finally, we need to be aware that we deal with mass spectral signals of peptides and not of full proteins. To infer the abundance or sequence of the whole protein requires additional computational steps which we will describe in the later part of this commentary..

2.2. Chromatographic Separation

After enzymatic digestion, we have obtained a mixture of peptides and maybe other sample compounds. If we would inject this sample directly into a mass spectrometer, we would obtain just one crowded mass spectrum containing many, and often overlapping, signals of all sample compounds. To alleviate this effect, we perform a simplification of the sample using chromatography as a first step.

The basic principle of chromatography is straightforward: sample molecules traverse the length of the chromatographic column and are retarded by chemical or physical interactions with the column material, the stationary phase. The amount of retardation depends on the nature of the molecules, the stationary phase and the solvent employed. The time at which a compound elutes is called the *retention time*, abbreviated rt. For peptides, the retention time is heavily influenced by their amino acid sequence. Peptides with similar amino acid compositions will elute at similar retention times.

To summarize, the peptides and all other sample components are separated by liquid chromatography and only after this separation, they are injected into the mass spectrometer. This leads to less complex subsets of the sample and to less crowded mass spectra. The drawback is that the data size increases and that we will obtain many mass spectra per sample, often several thousand instead of just one. Each spectrum represents a subset of the sample compounds as they elute from the column. As stated above, the set of all spectra obtained from a sample using one combination of column and mass spectrometer is the *LC-MS map*.

2.3. Mass Spectrometry Instrumentation

After the chromatographic separation, the sample molecules are injected into the mass spectrometer. It consists of an *ion source*, a *mass analyzer* and a *detector*. Electrospray Ionization (ESI) and Matrix-Assisted Laser Desorption/Ionization (MALDI) are the two most commonly employed ion sources for proteins and peptides.

2.3.1. Ion Sources

Electrospray Ionization (ESI) ionizes the analytes out of an aqueous solution and is therefore the method of choice for LC-MS setups. The liquid is pushed through a small capillary to which a potential difference is applied [3]. This causes a strong electric field which in turn leads to a clustering of charged molecules at the surface of the liquid. The equal charges repel each other and the liquid is pushed out of the capillary. It forms an aerosol, a mist of of small liquid droplets. The solvent molecules, which are usually more volatile than the analyte, evaporate and force the analyte molecules into closer vicinity. The molecules repel each other and break up the droplets. The process repeats until all solvent molecules are removed and the analytes form lone ions. Interestingly, even though this method was invented 19 years ago, and its underlying principles are even older, there is still an ongoing debate on the exact mechanisms of this ionization process [1].

ESI is well suited for complex samples and high-throughput experiments. Consequently, this text focuses on data from ESI ion sources and the resulting computational problems. From a data analysis point of view, it is important to realize that using ESI, copies of the same peptide can receive different charge states, i.e. the same peptide can appear with charge 2, 3 and 4, just to give an example. This complicates the analysis.

An ESI source can be operated in positive or negative ion mode. In positive ion mode, the compounds receive a proton as charged adduct, in negative mode they receive an electron. For tryptic peptides, most laboratories perform ESI mass spectrometry in positive ion mode, since the tryptic digest leads to peptides with negatively charged terminal amino acids. More generally, the mass-to-charge ratio (m/z) of a sample compound in ESI mass spectrometry is given by

$$\frac{m}{z} = \frac{M + im_a}{ie} \tag{1}$$

where M is the parent mass (the mass of the peptide), i is the number of charges and e is the elementary charge. m_a is the mass of the charged adduct. In most cases, this adduct is either an electron or a proton, but other adducts such as sodium (Na^+) or potassium K^+ are possible.

From a data analysis point of view, it is also important to realize that the ionization process is not perfect. Due to *competitive ionization*, some molecules are easily ionized whereas others do not ionize at all or cannot even be solved in liquid. These and other effects will cause the loss of some molecules before they even reach the detector. We need to keep in mind that using LC-MS we will only observe a subset of all sample compounds.

Finally, matrix-assisted laser desorption/ionization (MALDI) is a method to ionize molecules out of a dry, crystalline matrix via laser pulses. It is widely used to identify peptides in mixtures and but is not well-suited for quantitative measurements [4] and is thus not a focus of this text.

2.3.2. The Mass Analyzer

The mass analyzer measures the mass-to-charge ratio (m/z) of the ionized analyte. For our purposes, its key parameters are sensitivity, mass resolution and mass accuracy [4]. The sensitivity characterizes the ability of the mass analyzer to detect weak signals. Mass resolution and mass accuracy describe how well the analyzer is able to resolve signals with similar mass and how accurately it measures this mass, respectively. The mass resolution is a dimensionless unit and is expressed as the ratio of the mass of a signal peak in a mass spectrum and its Full-Width-At-Half-Maximum (FWHM).

Unfortunately, this is not a formal definition, but only a convention. The mass resolution is sometimes also calculated using the peak width at certain intensity thresholds or by measuring the intensity in a valley between two peaks. We will restrict ourselves to the first definition for the remainder of this text.

Finally, the mass accuracy is defined as the observed difference between the real mass of an analyte and the mass measured in the spectrum :

$$\text{mass accuracy} = \text{mass}_{\text{real}} - \text{mass}_{\text{measured}}. \tag{2}$$

The mass accuracy is often expressed in parts per million (ppm) :

$$\text{ppm} = 10^6 \times \text{mass accuracy}/\text{mass}_{\text{measured}}. \tag{3}$$

To give an example, if we know that the true mass of a compound is 1000.0 m/z and our mass spectrometer measures a signal for this compound at 999.99 m/z, then the accuracy of this measurement was ≈ 10 ppm. [5] suggested that this definition of mass accuracy is actually a mass deviation but the use of the term accuracy is wide spread.

There are four basic types of mass analyzer currently used in proteomics research. These are the Time-Of-Flight (TOF), Quadrupole, Fourier Transform Ion Cyclotron (FT-ICR) and the Orbitrap mass analyzer, each with its own strengths and weaknesses.

The TOF analyzer relies on a simple physical principle. It uses an electric field to accelerate the ions and then measures the time t they take to reach the detector. The ions with charge z are accelerated by a potential V. In this case, it holds that

$$t^2 = \frac{m}{z} \cdot \left(\frac{d^2}{2eV}\right) \tag{4}$$

where d is the length of the flight chamber and e is the elementary charge [1]. Thus, the mass-to-charge ratio m/z of an ion stands in a quadratic relationship to the flight time t and can readily be calculated.

This first mass spectrometers used TOF mass analyzers [6], and they are nowadays a well-established technology, reliable and inexpensive. Nevertheless, new methods have been invented in recent years. Therefore, we outline the basics of other mass analyzers as well.

The Quadrupole analyzer is a mass filter: it is capable of transmitting only ions with a specified m/z. It consists of four parallel electrodes. Each pair of opposing electrodes is connected electrically and a radio frequency voltage is applied between both pairs of electrodes. A direct current voltage is superimposed on the radio frequency voltage. After leaving the ion source, ions fly through the Quadrupole in between the electrodes. Only ions with a certain m/z will have a stable trajectory and can pass through. This allows selection and trapping of ions with a particular m/z but also to scan over a m/z range by changing the voltages. As an example, in the Quadrupole Ion Trap or Paul Trap, the ions are trapped for a certain time interval. Its basics were firstly described by [7]. Depending on the radio-frequency voltage applied, ions of a specific m/z can be trapped in the space between the electrodes. After that, the strength of the electric field is changed such that the ions are catapulted out of the trap and have their m/z recorded. Ion traps have the reputation of being very robust and are inexpensive [8]. They produced much of the proteomics data which is available today. Their disadvantage is the low mass accuracy: the trap can only capture a limited number of ions before they distort the accuracy of the mass measurement.

Recent developments try to circumvent this problem by changing size and shape of the ion trap, such as in the two-dimensional or linear ion trap, abbreviated as LTQ ("Linear Trap Quadrupole"). The Quadrupole can also be combined with a Time-Of-Flight analyzer, this combination is called QTOF (sometimes also written qTOF). In this setting, the quadrupole serves as an ion guide for the TOF analyzer. This results in higher sensitivity and mass accuracy for the TOF instrument [9].

The Fourier Transform-Ion Cyclotron (FT-ICR) instrument is also a trapping mass spectrometer. FT-ICR was invented by [10]. It determines the mass-to-charge ratio of ions based on the cyclotron frequency of the ions in a fixed magnetic field. This field is created under high vacuum in a Pienning ion trap. Once they are in the trap, the ions are excited to a cyclotron radius by an oscillating electric field perpendicular to the magnetic field in the trap. The mass-to-charge ratio determines the cycling frequency of an ion which is measured by detectors at fixed positions in the trap. The strengths of an FT-ICR mass analyzer are its high sensitivity, mass accuracy (1-2 ppm) and resolution. Recent developments include the combination of a Fourier Transform mass spectrometer with a Linear Trap Quadrupole [11].

The LTQ Orbitrap is the most recently developed analyzer [12]. The Orbitrap operates by trapping ions around a central spindle electrode. The suggestive name derives from the ions traveling in an orbit around the central electrode in the trap. An outer barrel-like electrode is placed coaxially with the central electrode. The ions are tangentially injected into the trap and oscillate around the central electrode. In addition, they also move back and forth along the axis of the electrode. Their m/z values are calculated from the frequency of their oscillations by applying the Fourier Transform. Orbitraps have a mass accuracy and resolution comparable to FT-ICR instruments but at a fraction of the costs [13]. Consequently, more and more laboratories use them for their experiments.

2.3.3. The Mass Detector

Finally, the mass detector counts the number of ions at each m/z. Possible detectors are *photographic plates*, *faraday cylinders* or *array detectors* [1]. Most detectors need some time to recover after an ion hit. This time span is called *dead time* [9]. This means that ions with very similar m/z might not be accurately recorded.

3. Algorithms for Quantitative LC-MS Experiments

We spent the first half of this commentary to explain to technical setup of an LC-MS experiment. In its second part, we will introduce the key computational steps in a typical LC-MS data analysis workflow, starting with low-level filtering denoising of the mass spectra and ending with higher-level analysis such as the alignment of retention times and quantification of sample compounds.

3.1. Low-level Signal Processing and Peptide Feature Discovery

The data analysis starts with the unprocessed (raw) LC-MS map. Since the map is subject to background chemical and electronic noise, together with systemic contaminants in the mobile phase in the chromatographic column, methods for noise reduction and signal enhancement are commonly applied. Typical low-level processing steps include smoothing using median, moving average or the or the more sophisticated Savitzky-Golay filter [14].

In some cases, the LC-MS map will also contain noise of lower frequency, usually referred to as baseline noise. Methods that are applied to estimate and then subtract the baseline from a mass spectrum are iterative local regression [15], matched filter [16] and Fourier transformations [17].

The next step involves the detection and quantification of peptide signals in the LC-MS map. The computational challenge is to deal with data from different instruments, to recover weak peptide signals and to assign abundance and charge state correctly.

3.2. Computational Quantification of Peptides and Proteins

There are several algorithms for the detection of peptide features in mass spectra. The first step is usually a *seeding* step: the algorithm tries to identify prominent peaks in the map that could be part of a peptide signal. Feature detection algorithms mostly differ in the criteria they use to select these seeds. After this, most algorithms separate the seeding points from the surrounding background and compute a preliminary bounding box.

After this, all algorithms perform a *refinement* step: all data points in the previously computed bounding box are compared to an averagine model. Most algorithms do not take the elution behavior of the isotope peaks into consideration and some even apply only a crude approximation to the true isotope pattern. Quantification algorithms also differ in the mass resolution and accuracy they expect: some require highly resolved spectra, others perform well only on low resolution data. In the remainder of this section, we will review the state of the art of LC-MS feature detection algorithms. We will however take only algorithms into consideration that are freely available for academic purposes i.e. no commercial software or algorithms for which no free implementation exists.

3.2.1. Seeding

There are several possible criteria for seed selection. A straightforward approach is to sort all data points in the LC-MS map and to check all points above a given intensity threshold [18]. This approach has also been coined as *intensity descent* [19]. This strategy is straightforward to implement but becomes prohibitively slow if the LC-MS map is complex and contains many noise signals. In this case, an intensity-based seeding method will select many noise signals or non-peptidic signals as feature candidates and perform many unnecessary refinement steps or even require manual inspection to yield good results. Thus, it makes sense to take more criteria than the intensity of a signal into consideration. [20] compute various higher-order statistics for each seed, such as signal-to-noise ratio, consistency across several spectra and intensity difference to neighboring points. Only seeds that pass user-defined thresholds in all these criteria are considered valid seeds. The approach by [19] is similar in spirit, but in contrast to [20] they interleave refinement and seeding step. In short, they check groups of high-intensity points for correlation with an averagine model first, remove them from the spectrum and only then proceed with lower intensity signals. This leads to fewer false positives. [21] also perform an intensity descent but also take more sophisticated peak characteristics into account. First of all, their algorithm does not operate on data points but groups points into peaks and only considers peaks of a user-defined minimum width. The algorithm also considers only peaks that occur at consistent m/z positions across at least 5 spectra. It iteratively groups adjacent peaks of high-intensity into a cluster, fits an averagine model for refinement and then proceeds until all peaks are processed.

An LC-MS map can be represented as a two-dimensional image if we interpret m/z and retention time as the two dimensions and represent the intensity by a color code. It is therefore an evident idea to use image processing methods for LC-MS feature detection. [22] apply the Watershed transformation [23] which is another intensity-based seeding approach. This transformation performs a segmentation of the LC-MS map. The resulting segments are disjoint and homogeneous regions in the map. The Watershed transformation comes from the field of digital image processing and is usually defined in terms of discrete and equally spaced points. Consequently, [22] resample the LC-MS map to achieve a rectangular grid. In any case, this transformation is fast but the resampling might decrease the resolution of the data. Furthermore, it does not take the shape of the signal into consideration which is an important and reliable information in the case of isotope pattern.

The THRASH (Thorough High-Resolution Analysis of Spectra by Horn) algorithm [24] uses a seeding approach based on a histogram of the signal intensities. Apart from that, this algorithm is similar to the previous ones. But it is noteworthy since it was the first published LC-MS feature detection algorithm. Unfortunately, its first implementation is not longer available. The software DECON2LS offers a re-implementation and is available from the NCRR Proteomics Resource at the Pacific Northwest National Laboratory, Richland, US.[1]. The THRASH algorithm was also re-implemented and modified by [25]. This approach assumes that the background signal over a given m/z range will be acquired more frequently than the signal of the isotope peaks. It builds a histogram of the signal intensities for a small m/z range, around 25 Th. The background intensity is estimated as the intensity with the highest frequency. The noise is estimated by Full-Width-at-Half-Maximum (FWHM) of the smoothed histogram, and the signal-to-noise ratio (S/N) is calculated for each signal point (I_p) using

$$S/N = \frac{I_p - I_b}{\text{FWHM}}$$

where I_b is the background intensity. Peaks surpassing a user-defined S/N threshold (usually around $2 - 3$) are chosen as seeds.

The software tools MSINSPECT [26] and SPECARRAY [27] both use the wavelet transformation to filter the raw spectra and later search for local-maxima to identify peaks. The algorithms differ in the type of mother wavelet they use. [26] apply the Marr-Wavelet (also called Mexican Hat) whereas [27] use the Symmlet8-wavelet. The wavelets help to suppress signals without the typical shape of a MS peak and thus reduce the number of seeds to be considered. Unfortunately, the respective publications do not give further details on the feature discovery process.

The peptide quantification algorithm SWEEPWAVELET [28, 29] is part of the OpenMS software library [30]. It also implements a wavelet-based feature detection algorithm. In contrast to other quantification algorithms, it uses a novel wavelet function that mimicks the peak intensity pattern of peptides in mass spectra.

Finally, the software SUPERHIRN [31] implements a fast feature detection routine which is an improved version of the pattern matching method previously published in [32]. In short, the algorithm starts by fitting an averagine template to peaks with high intensity in each spectrum. Peptide features are identified by searching for local minima in the error function, i.e., in the squared absolute distance between averagine template and spectrum. They use several heuristics

[1] http://ncrr.pnl.gov/software/

to improve the running time of their algorithm. As an example, they compute the error function for larger bins of spectra and region with low noise content first. After well-resolved features are found, they remove the corresponding peaks from the spectra and re-compute the error function to find lower intensity and overlapping features.

3.3. Refinement

All these approaches result in a set of potentially interesting signals in the LC-MS map which might be caused by a peptide. In a next step, these seeds are filtered for false positives. To do so, we fit a theoretical model of a peptide signal to the peaks in the neighborhood of each seed and determine the goodness of this fit, usually using a measure of correlation of peak intensities predicted by the model and the data. Signals with a poor fit to the theoretical predictions are discarded. Almost all algorithms introduced above use the averagine model as introduced by [33]. Current approaches mainly differ in the implementation and the method of charge determination. Table 3.3. gives an overview. In the following, we will highlight these differences.

Name	Reference	Seeding strategy	Charge determination
AID-MS	[19]	intensity descent	interpeak spacing
HARDKLOER	[25]	histogram	m/z distance
LCMS2D	[21]	intensity descent	variable selection
MAPQUANT	[22]	watershed	averagine model
MSINSPECT	[26]	Marr Wavelet	averagine model
MZMINE	[34]	intensity descent	averagine model
SPECARRAY	[27]	Symmlet8 Wavelet	m/z distance
SUPERHIRN	[31]	error function	averagine model
SWEEPWAVELET	[28]	Isotope Wavelet	2D peptide signal model
THRASH	[24]	histogram	FT + Patterson

Figure 1. *Overview of quantification algorithms including references.*

Noteworthy is the software MAPQUANT since it includes a parameter of the peak shape into its averagine model. This tool models the peak shape using a Gaussian distribution. By changing the peak widths, these algorithms can account for different mass resolutions. Unfortunately, MAPQUANT reads only proprietary file formats and is apparently not longer maintained. All other algorithms do not take the peak shape into consideration but reduce the raw data points to peaks and compare peak intensities to averagine intensities.

SWEEPWAVELET uses a two-dimensional peptide signal model that takes both m/z pattern and elution behaviour of the peptide into consideration. This filter is used to remove false positive signals, signals caused by contaminants and other non-peptide compounds. It also makes efficient use of the redundancy in mass spectra to yield accurate estimates of mass, charge and the centroid of retention time.

An interesting feature of HARDKLOER is its ability to model non-standard isotope pattern. The user can define averagine amino acids with non-standard composition to account for modified or isotope-enriched peptides. Obviously, this makes only sense for high-resolution data. HARDKLOER uses a simplified averagine model. Their model is based on a Poisson distribution which is fast to compute but results in a less accurate model [26].

3.4. LC-MS Map Alignment

After the step of peptide feature detection and quantification, we know mass, charge and retention time of the peptides in the LC-MS map. We estimate their relative abundance from the signal intensity. But the comparison of these abundances across different samples is not straightforward. It is hampered by the fact that the retention time at which a peptide elutes from the chromatographic column is not stable across different experiments. The elution behavior is distorted in complex ways by differences in column performance due to changes in ambient pressure and temperature.

Several algorithms were developed to correct these distortions. Examples are hidden Markov models [35], dynamic time warping [14], or clustering [16]. These algorithms usually make different assumptions about the type of distortions (e.g. linear or non-linear) [36]. Some of them can be applied to the raw spectra directly and others operate on the m/z and rt coordinates of peptide features.

The decharging of the feature coordinates is a step that can be performed after the alignment of retention times. The peptide features with mass-to-charge ratio m/z are simply mapped to the corresponding parent peptide mass and have their intensity measurements summarized. Obviously, this requires a correct charge estimate. The process of decharging can be a non-trivial step for complex samples.

3.5. Statistical Analysis

The actual detection of significant differences in peptide abundances is the next step. It involves the comparison of peptide signal intensities between the aligned LC-MS maps. The resulting statistical problems are comparable to issues in microarray data analysis and similar methods, such a linear models [37] or lowess smoothing for normalization [16], are applied. Further computational steps include the mapping of peptide sequences obtained from MS/MS to features as well as the abundance estimation on the protein level.

3.6. Peptide and Protein Sequencing

From the data generated in an LC-MS experiment, we can determine the mass, charge and retention time of the peptides contained in the sample under examination. But for most applications, this is not sufficient and we are also interested in the actual amino acid sequences. This requires an additional measurement, the acquisition of MS/MS or tandem MS spectra. We will briefly explain the principle of peptide sequencing using tandem mass spectrometry since we will refer to this technology from time to time in the later chapters of this thesis.

An MS/MS spectrum provides additional information for selected signals in the LC-MS map. The MS/MS spectrum is a mass spectrum and contains m/z and intensity measurements of fragments of the precursor ion. The mass spectrometer usually records MS/MS spectra automatically for signals with an intensity higher than a given threshold. This procedure is called *Data-dependant Acquisition* (DDA) and the compounds selected for MS/MS recording are called *precursor ions*.

There are different approaches to perform this fragmentation. The most frequently used method is *Collision Induced Dissociation* (CID) [38], also called *Collisionally Activated Dissociation* (CAD). In short, the mass spectrometer captures selected ions, the *parent* or *precursor ions*, accelerates them using an electric potential and allows them to collide with neutral gas molecules, such as helium or argon. This results in a breakage of the parent ion into smaller fragments. These fragments are recorded in a MS/MS spectrum. Recently, new fragmentation methods have emerged such as *Electron Transfer Dissociation* (ETD) or *Electron Capture Dissociation* (ECD) [39, 40].

If peptides are subjected to CID, they usually break at the peptide bonds. The mass spectrum records the m/z values of the fragment masses in the MS/MS spectrum. The mass differences between consecutive fragments match the corresponding amino acids. In many cases, this allows at least a partial reconstruction of the amino acid sequence (*de-novo sequencing*). If this is not possible due to noise, missing fragments or unexpected post-translational modifications, we can compare the tandem spectrum against a library of manually annotated spectra(*spectral library searching*) [41]. Another possibility is to compute theoretical MS/MS spectra for peptides obtained from a protein sequence database and to compare them against the experimental spectrum (*sequence database searching*). The latter method is the most common one. Popular algorithms are, for example, Sequest [42], Mascot [43] or InsPecT [44].

This is of course only a rough sketch of a complex process. But for the remainder of this thesis, it suffices to know that we can, in theory, obtain the amino acid sequences for the peptides in our sample. This is a difficult task, since peptide identification using MS/MS is error-prone and the algorithms available today are known to produce high false positive rates. We have to be aware that some of the obtained amino acid sequences will be wrong, others might only partially be correct.

It is also important to realize that most MS/MS data available today was acquired in an automatic fashion, using the abovementioned Data-dependent Acquisition. In this case, the mass spectrometer selects prominent signals in each spectrum automatically for fragmentation. This usually leads to an undersampling of peptide ions [45] since weak signals will be missed. It is almost impossible to sequence every peptide in a sample in a single-pass analysis.

4. Summary and Conclusions

We summarized the computational challenges in the analysis of data from a liquid chromatography-mass spectrometry LC-MS experiment, highlighted some algorithms to address them and reviewed mass spectrometry instrumentation. Mass spectrometry has revolutionized the way we can examine the protein contents of the cell and the wealth of information we accumulated in recent years is breathtaking. At the same time, new MS technologies such as MALDI imaging [46] are emerging which will require new algorithms and methodologies for

Nevertheless, as the speed of data acquisition increases, there remain significant bottlenecks and unsolved problems in LC-MS data analysis. Topics such as the accurate estimation of false discovery rates for MS database searches [47], interleaving of quantification and alignment of LC-MS maps [48], proteogenomic annoation [49] as well as the accurate identification of protein modification are actively being researched and pose new challenges for Computational Scientists interested in the field of proteomics.

References

[1] Edmond de Hoffmann and Vincent Stroobant. *Mass Spectrometry: Principles and Applications*. Wiley & Sons, 2004.

[2] M.J. MacCoss and D. E. Matthews. Quantitative MS for proteomics: Teaching a new dog old tricks. *Anal. Chem.*, 77(15):294A–302A., 2005.

[3] JB Fenn, M Mann, CK Meng, SF Wong, and CM Whitehouse. Electrospray ionization for mass spectrometry of large biomolecules. *Science*, 246(4926):64–71, 1989.

[4] M. Mann and R. Aebersold. Mass spectrometry-based proteomics. *Nature 422*, 422:198 – 207, 2003.

[5] Roman Zubarev and Matthias Mann. On the Proper Use of Mass Accuracy in Proteomics. *Mol Cell Proteomics*, 6(3):377–381, 2007.

[6] W. C. Wiley and I. H. McLaren. Time-of-Flight Mass Spectrometer with Improved Resolution. *Review of Scientific Instruments*, 26(12):1150–1157, 1955.

[7] W. Paul and H. Steinwedel. Ein neues massenspektrometer ohne magnetfeld. *Zeitschrift fr Naturforschung A*, 8(7):448–450, 1953.

[8] Raymond E. March. Quadrupole ion trap mass spectrometry: a view at the turn of the century. *International Journal of Mass Spectrometry*, 200:85–312, 2000.

[9] Igor V. Chernushevich, Alexander V. Loboda, and Bruce A. Thomson. An introduction to quadrupole-time-of-flight mass spectrometry. *Journal of Mass Spectrometry*, 36(8):849–865, 2001.

[10] Melvin B. Comisarow and Alan G. Marshall. Fourier transform ion cyclotron resonance spectroscopy. *Chemical Physics Letters*, 25:282–283, 1974.

[11] J.E.P. Syka, J.A. Marto, D.L. Bai, S. Horning, M.W. Senko, J.C. Schwartz, B. Ueberheide, B. Garcia, S. Busby, T. Muratore, J. Shabanowitz, and D.F. Hunt. Novel linear quadrupole ion trap/ft mass spectrometer: Performance characterization and use in the comparative analysis of histone h3 post-translational modifications. *Journal of Proteome Research*, 3(3):621–626, 2004.

[12] Qizhi Hu, Robert J. Noll, Hongyan Li, Alexander Makarov, Mark Hardman, and R. Graham Cooks. The orbitrap: a new mass spectrometer. *Journal of Mass Spectrometry*, 40(4):430–443, 2005.

[13] Michaela Scigelova and Alexander Makarov. Orbitrap mass analyzer - overview and applications in proteomics. *Proteomics*, 6(S2):16–21, 2006.

[14] W. Wang, H. Zhou, H. Lin, S. Roy, T.A. Shaler, L.R. Hill, S. Norton, P. Kumar, M. Anderle, and C.H. Becker. Quantification of proteins and metabolites by mass spectrometry without isotopic labeling or spiked standards. *Analytical Chemistry*, 75(18):4818–4826, 2003.

[15] Michael Wagner, Dayanand Naik, and Alex Pothen. Protocols for disease classification from mass spectrometry data. *Proteomics*, 3(9):1692–1698, 2003.

[16] A. Sauve and T. Speed. Normalization, baseline correction and alignment of high-throughput mass spectrometry data. In *Proceedings Gensips*, 2004.

[17] K. A. Baggerly, J. S. Morris, J. Wang, D. Gold, L. C. Xiao, and K. R. Coombes. A comprehensive approach to the analysis of matrix-assisted laser desorption/ionization-time of flight proteomics spectra from serum samples. *Protemics*, 3:1667–1672, 2003.

[18] C. Gröpl, E. Lange, K. Reinert, O. Kohlbacher, M. Sturm, C. G. Huber, B. Mayr, and C. Klein. Algorithms for the automated absolute quantification of diagnostic markersin complex proteomics samples. In Michael Berthold, editor, *Procceedings of CompLife 2005*, Lecture Notes in Bioinformatics, pages 151–163. Springer, Heidelberg, 2005.

[19] L. Chen, S.K. Sze, and H. Yang. Automated intensity descent algorithm for interpretation of complex high-resolution mass spectra. *Analytical Chemistry*, 78(14):5006–5018, 2006.

[20] Mikko Katajamaa, Jarkko Miettinen, and Matej Orešič. MZmine: Toolbox for processing and visualization of mass spectrometry based molecular profile data. *Bioinformatics*, 22(5):634–636, 2006.

[21] Peicheng Du, Rajagopalan Sudha, Michael B. Prystowsky, and Ruth Hogue Angeletti. Data reduction of isotope-resolved lc-ms spectra. *Bioinformatics*, 23(11):1394–1400, 2007.

[22] K. C. Leptos, D. A. Sarracino, J. D. Jaffe, B. Krastins, and G. M. Church. MapQuant: Open-Source software for large-scale protein quantification. *Proteomics*, 6(6):1770–1782, 2006.

[23] Lee Vincent and Pierre Soille. Watersheds in digital spaces: An efficient algorithm based on immersion simulations. *IEEE PAMI, 1991*, 13(6):583–598, 1991.

[24] David M. Horn, Roman A. Zubarev, and Fred W. McLafferty. Automated reduction and interpretation of high resolution electrospray mass spectra of large molecules. *Journal of the American Society for Mass Spectrometry*, 11(4):320–332, April 2000.

[25] M.R. Hoopmann, G.L. Finney, and M.J. MacCoss. High-speed data reduction, feature detection, and ms/ms spectrum quality assessment of shotgun

proteomics data sets using high-resolution mass spectrometry. *Analytical Chemistry*, 79(15):5620–5632, 2007.

[26] M. Bellew, M. Coram, M. Fitzgibbon, M. Igra, T. Randolph, P. Wang, D. May, J. Eng, R. Fang, C.-W. Lin, J. Chen, D. Goodlett, J. Whiteaker, A. Paulovich, and M. McIntosh. A suite of algorithms for the comprehensive analysis of complex protein mixtures using high-resolution LC-MS. *Bioinformatics*, 22(15):1902–1909, 2006.

[27] Xiao-jun Li, Eugene C. Yi, Christopher J. Kemp, Hui Zhang, and Ruedi Aebersold. A Software Suite for the Generation and Comparison of Peptide Arrays from Sets of Data Collected by Liquid Chromatography-Mass Spectrometry. *Mol Cell Proteomics*, 4(9):1328–1340, 2005.

[28] Ole Schulz-Trieglaff, Rene Hussong, Clemens Gröpl, Andreas Hildebrandt, and Knut Reinert. A fast and accurate Algorithm for the Quantification of peptides from LC-MS data. In Terence P. Speed and Haiyan Huang, editors, *Research in Computational Molecular Biology, 11th Annual International Conference, RECOMB 2007, Oakland, CA, USA, April 21-25, 2007*, volume 4453 of *Lecture Notes in Computer Science*, pages 473–487. Springer, 2007.

[29] Ole Schulz-Trieglaff, Rene Hussong, Clemens Gröpl, Andreas Leinenbach, Andreas Hildebrandt, Christian Huber, and Knut Reinert. Computational Quantification of Peptides from LC-MS data. *Journal of Computational Biology*, 2008.

[30] Marc Sturm, Andreas Bertsch, Clemens Groepl, Andreas Hildebrandt, Rene Hussong, Eva Lange, Nico Pfeifer, Ole Schulz-Trieglaff, Alexandra Zerck, Knut Reinert, and Oliver Kohlbacher. Openms - an open-source software framework for mass spectrometry. *BMC Bioinformatics*, 9, 2008.

[31] Lukas N Mueller, Oliver Rinner, Alexander Schmidt, Simon Letarte, Bernd Bodenmiller, Mi-Youn Brusniak, Olga Vitek, Ruedi Aebersold, and Markus Müller. Superhirn - a novel tool for high resolution lc-ms-based peptide/protein profiling. *Proteomics*, 7(19):3470–3480, Oct 2007.

[32] Steven Gay, Pierre-Alain Binz, Denis F. Hochstrasser, and Ron D. Appel. Peptide mass fingerprinting peak intensity prediction: Extracting knowledge from spectra. *Proteomics*, 2(10):1374–1391, 2002.

[33] M.W. Senko, S.C. Beu, and F.W. McLafferty. Determination of monoisotopic masses and ion populations for large biomolecules from resolved isotopic distributions. *Journal of the American Society for Mass Spectrometry*, 6:229–233, 1995.

[34] Mikko Katajamaa and Matej Orešič. Processing methods for differential analysis of LC/MS profile data. *BMC Bioinformatics*, 6:179, 2005.

[35] Jennifer Listgarten, Radford M. Neal, Sam T. Roweis, Peter Wong, and Andrew Emili. Difference detection in lc-ms data for protein biomarker discovery. *Bioinformatics*, 23:e198–e204, 2007.

[36] Katharina Podwojski, Arno Fritsch, Daniel C. Chamrad, Wolfgang Paul, Barbara Sitek, Kai Stuhler, Petra Mutzel, Christian Stephan, Helmut E. Meyer, Wolfgang Urfer, Katja Ickstadt, and Jorg Rahnenfuhrer. Retention time alignment algorithms for LC/MS data must consider non-linear shifts. *Bioinformatics*, 25(6):758–764, 2009.

[37] Ann L. Oberg, Douglas W. Mahoney, Jeanette E. Eckel-Passow, Christopher J. Malone, Russell D. Wolfinger, Elizabeth G. Hill, Leslie T. Cooper, Oyere K. Onuma, Craig Spiro, Terry M. Therneau, and H. Robert Bergen, III. Statistical analysis of relative labeled mass spectrometry data from complex samples using anova. *Journal of Proteome Research*, 7(1):225–233, 2008.

[38] J. Mitchell Wells and Scott A. McLuckey. Collision[hyphen (true graphic)]induced dissociation (cid) of peptides and proteins. In A. L. Burlingame, editor, *Biological Mass Spectrometry*, volume Volume 402, pages 148–185. Academic Press, 2005.

[39] John E. P. Syka, Joshua J. Coon, Melanie J. Schroeder, Jeffrey Shabanowitz, and Donald F. Hunt. Peptide and protein sequence analysis by electron transfer dissociation mass spectrometry. *Proceedings of the National Academy of Sciences*, 101(26):9528–9533, 2004.

[40] R.A. Zubarev, N.L. Kelleher, and F.W. McLafferty. Electron capture dissociation of multiply charged protein cations. a nonergodic process. *Journal of the American Chemical Society*, 120(13):3265–3266, 1998.

[41] Henry Lam, Eric W. Deutsch, James S. Eddes, Jimmy K. Eng, Nichole King, Stephen E. Stein, and Ruedi Aebersold. Development and validation of a spectral library searching method for peptide identification from ms/ms. *Proteomics*, 7(5):655–667, 2007.

[42] JR 3rd Yates, JK Eng, AL McCormack, and Schieltz D. Method to correlate tandem mass spectra of modified peptides to amino acid sequences in the protein database. *Anal Chem.*, 15:1426–1436, 1995.

[43] D. N. Perkins, D. J. C. Pappin, D. M. Creasy, and J. S. Cottrell. Probability-based protein identification by searching sequence databases using mass spectrometry data. *Electrophoresis*, 20:3551–3567, 1999.

[44] S. Tanner, H. Shu, A. Frank, L.-C. Wang, E. Zandi, M. Mumby, P. A. Pevzner, and V. Bafna. Inspect: Fast and accurate identification of post-translationally modified peptides from tandem mass spectra. *Anal. Chem.*, 77(14):4626–39, 2005.

[45] H. Liu, R.G. Sadygov, and J.R. Yates. A model for random sampling and estimation of relative protein abundance in shotgun proteomics. *Analytical Chemistry*, 76(14):4193–4201, 2004.

[46] Liam A. McDonnell and Ron M.A. Heeren. Imaging mass spectrometry. *Mass Spectrometry Reviews*, 26(4):606–643, 2007.

[47] Nitin Gupta and Pavel A. Pevzner. False discovery rates of protein identifications: A strike against the two-peptide rule. *Journal of Proteome Research*, 8(9):4173–4181, September 2009.

[48] Frank Suits, Jorge Lepre, Peicheng Du, Rainer Bischoff, and Peter Horvatovich. Two-dimensional method for time aligning liquid chromatographymass spectrometry data. *Analytical Chemistry*, 80(9):3095–3104, May 2008.

[49] Charles Ansong, Samuel O. Purvine, Joshua N. Adkins, Mary S. Lipton, and Richard D. Smith. Proteogenomics: needs and roles to be filled by proteomics in genome annotation. *Brief Funct Genomic Proteomic*, page eln010, 2008.

In: Proteomics
Editor: Giselle C. Rancourt

ISBN: 978-1-61668-691-8

Chapter 5

Method for Prediction of Protein-Protein Interactions in Yeast using Genomics/Proteomics Information and Feature Selection

J.M. Urquiza, I. Rojas, H. Pomares, L. J. Herrera*
Department of Computer Architecture and Computer Technology,
University of Granada, 18017 Granada, Spain

Abstract

Nowadays, one of the most important goals of Proteomics is the prediction of protein-protein interactions (PPIs), whose knowledge is vital for all biological processes. In the present paper we propose an approach to the prediction of protein-protein interactions in yeast based on the well-known paradigm of Support Vector Machines (SVM) for classification and feature selection methods using Genomics/Proteomics information from the main databases. In order to obtain higher values of specificity and sensitivity for our predictions, we took a high reliable set of positive and negative examples for the construction of the SVM model. We then extracted a set of proteomic/genomic features from these examples and also introduced a similarity measure in the calculation of the features,

*E-mail address: (jurquiza, irojas, hpomares, jherrera)@atc.ugr.es

that allows us to improve the prediction capability of our model. In the analysis of the results, we also applied our approach to in vitro datasets, obtaining high accuracy classifications. Our final SVM classifiers obtain a low error rate in the prediction for each pair of proteins of several datasets for both in vitro and in silico methodologies.

Keywords: Support Vector Machine, Protein-Protein Interaction, Prediction, Feature Selection, Genomics/Proteomics Databases.

1. Introduction

In current Proteomics, the elucidation of the structure, function and interactions of proteins that constitute cells and organisms [1] has taken an important role in the researching efforts. In fact, protein-protein interaction (PPI) networks contain a large amount of useful information for the functional characterization of proteins and promote the understanding of the complex molecular relationships that determine the phenotype of a cell [2]. Most fundamental biological processes involve protein-protein interaction [3]. Hence, a comprehensive determination of all PPIs that can take place in a cell would be an invaluable asset for the understanding of biology at the systems level [2] providing a powerful tool for studying diseases and their corresponding drug targets (for example neuro-degenerative diseases such as Parkinson, Alzheimer, etc. See figure 1).

The experimental data on protein-protein interactions are obtained under different conditions, for different organisms and present a wide range of details. For example, Y2H is widely used to make it available the identification of interacting proteins, electron microscopy provides relative positional information of interacting proteins or protein subunits, and crystallography provides full atomic detail of interaction surfaces.

High-throughput experiments can be applied on a genome-wide scale to identify protein interaction partners while more specialized biophysical methods can characterize interactions in terms of kinetics, dynamics and mechanics of binding processes.

In addition to direct detection of physical protein interactions, computational methods of protein interaction prediction are becoming more powerful in their ability to identify potential protein interaction partners, specific interacting

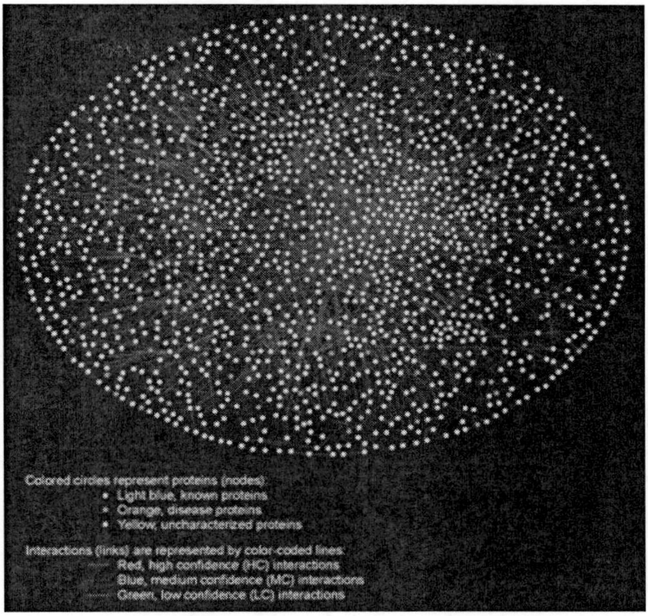

Figure 1. A human protein-protein interaction network: the first step toward a human interactome map (Figure reproduced from Cell DOI: 10.1016/j.cell.2005.08.029.)

protein domains and indirect functional associations between proteins.

In order to analyze interactions in Proteomics, there exist two complementary branches. On the one hand, the analysis of protein complexes using affinity purification followed by mass spectrometry (AP/MS) is capable of identifying directly and indirectly associated proteins. On the other hand, the already mentioned high-throughput yeast two-hybrid (Y2H) analysis, which identifies direct, binary interactions [4]. Nowadays, these methodologies provide interactions for several organisms such as *S. cerevisiae* [5–8], *C.elegans* [9, 10], *D. melanogaster* [11,12], including *H.sapiens* [13], and more are forthcoming [14].

S. cerevisiae is the most widely analyzed organism even though its interactome is still far to be complete [14, 15]. Besides, it is thought that data obtained from high-throughput (HT) experiments have a high unknown number of false positives, i.e., interactions that are spurious or biologically irrelevant and do not occur in the cell [14–17]. For some studies, the authors specify the reliable interactions as 'high confidence' or 'core' interactions [6, 9, 11], or assess the quality of the interactions using several sources or datasets [17–19].

In supervised classification methods, such as those used to train an SVM model, we need a positive and negative set of examples. The set of positive examples is called the Gold Standard Positive (GSP) set and the set of negative examples is called the Gold Standard Negative (GSN) set. This GSN interaction set is as important as the GSP sets. However, negative interactions are unfortunately not reported by experimentalists [14]. There are several approaches to compile negative sets. Typically, negative sets are composed of pairs of proteins that do not share the same cellular compartment or that were randomly paired [3, 20]. These approaches could introduce a bias in the precision. In [21] the authors used a semantic similarity based on the annotations in Gene Ontologies [22] to predict negative interactions. In contrast to these approaches, Saeed and Deane [14] provided in their work a high quality set of positive and negative examples. The GSN dataset was generated using a combination of protein features (function, cellular localization, expression) and lack of homologous interaction.

In this work, we design a new predictor of protein-protein interactions in S. cerevisiae (Yeast) based on Support Vector Machines (SVM). With the purpose of training the SVM models we have used a reliable set of positive and a subset of negative pairs taken from Saeed et al. [14] for yeast. Through a feature extraction process we have obtained a total of 26 genomic/proteomic features, introducing a similarity measure to calculate some features to improve the prediction power of our approach. Subsequently, we obtain the most relevant features by means of feature selection methods proposing a modified method derived from a combination of other classical feature selection methods. Using the selected set of features, we eventually construct SVM predictors to obtain a low classification error and high values of specificity and sensitivity in the prediction of PPIs. To verify the prediction power of our proposal, we use several experimental or computational obtained datasets from Yu et al. [23] and Saeed et al. [14]. Finally, due to the fact that binary interaction mapping has grown, there exist information that must be assessed for data quality. Given the importance of protein interactions and the demand for better and more comprehensive maps, standardized experimental methods for quality control are crucial [4]. For this reason, our SVM predictors additionally return a score for each PPI that permits to give a quality measure, and therefore our proposed approach could also be considered as a validation tool.

The rest of the paper is organized as follows. Section 2. describes how the features were extracted from genomic/proteomic databases. Section 3. summarizes the used variable selection approaches. Section 4. is a brief theoretical description of SVM for classification. Section 5. presents the results of this work using a number of different datasets from Yu et al. [23] and Saeed et al. [14]. Finally, some conclusions are drawn in section 6..

2. Feature Extraction

The dataset in this work has been obtained from Saeed et al. [14]. They provide a set of 4 million negative pairs of proteins and a set of 4809 interacting pairs of proteins for yeast. We have used all positive examples (4809) and a random subset of 4895 negative examples, giving a total of 9704 pairs.

From this dataset, we extracted 26 genomic/proteomic features for yeast from diverse databases. For each pair, we have mined the information from the following databases:

- SwissPfam (version 22.0) [24] supplies information on protein domains.
- Gene Ontology Annotation (GOA) Database [25] provides high quality annotation of Gene Ontology (GO) [22] (version May 2008). GOA classifies protein annotation into three ontologies: biological processes (P), cellular components (C) and molecular functions (F).
- MIPS Comprehensive Yeast Genome Database CYGD (version June 2008) [26] gathers information on molecular structure and functional network in Yeast. We consider all catalogs: functional, complexes, phenotype, proteins and sub-cellular compartments.
- HINTdb (version 13 June 2007) [27]. The Homologous Interactions (HINT) database is a collection of protein-protein interactions and their homologs in one or more species. Using this database, we have included

homology information for a given pair of proteins. Homology refers to any similarity between characteristics that is due to their shared ancestry.

- 3did database (version 25 May 2008) [28] gives a collection of domain-domain interactions in proteins whose 3D-structures are known in Protein Data Bank (PDB) [29].

The fundamental idea for our feature extraction approach is to find the number of common terms in a particular database or catalog between two proteins, as we consider that each protein has a number of associated terms in each database. Furthermore, we introduce a similarity measure that shows the percentage of common terms in a pair with respect to the sum of all terms which have two given proteins, for the databases used. This is a measure called "local" and can be represented as Eq. 1:

$$sim_{local} = \frac{n}{N} \tag{1}$$

where n is the number of shared elements between two proteins in a pair and N is the total number of elements in a pair of proteins.

A different measure is also proposed that shows the percentage of common terms in a pair with respect to the sum of all terms in a database. It will be refered to as "global similarity".

We present now the 26 features used for yeast, indicating between parenthesis the order and nature of the data (real or integer):

- Common GO annotation terms between two proteins of three ontologies (P,C,F) together (1st integer) and their "local" and "global" similarity measures (13th real, 14th real). Moreover we considered each separately ontology (5th P integer, 6th C integer, 7th F integer) and their respective "local similarity" measure (16th real, 17th real, 18th real) and "global ones" (19th real, 20th real, 21st real).

- Number of homologous between two proteins obtained from HintDB (2nd integer).

- Common Pfam domains between two proteins, extracted from SwissPfam, which are found in the 3did database (3rd integer). The similarity would be the number of common domains Pfam-3did between the total of Pfam domains that two proteins have (15th integer).

- mRNA co-expression value from Rosetta Compendium extracted from Jansen et al. work [3] (4th real).

- Considering each MIPS catalog separately, we counted the number of common terms (catalog identifier) between two proteins (functional 8th integer, complexes 9th integer, proteins 10th integer, phenotypes 11th integer, sub-cellular compartments 12th integer). Furthermore we considered their "local similarity" measures (22nd real, 23rd real, 24th real, 25th real, 26th real).

Therefore, in our feature extraction we have taken into account a set of features that covers a wide range of biological information about proteins. From mRNA co-expression to domain information. Research about protein structure suggests that the fundamental unit is a domain, and this kind of information is also used in other works [1, 3, 20] obtaining an improvement in the prediction and analysis of PPIs.

Consequently, we have included domain proteins data in our feature extraction approach. We have also considered expression profiles that were also used in other works [3, 30] to help the classification process. In addition, Saeed et al. [14] have demonstrated the helpful use of homologous information in order to construct datasets and improving the prediction accuracy, and thus we added data from the HintDB designed by Patil et al. [27] to cover this topic. Finally, we have considered GO annotations and MIPS CYGD catalogs that are widely used in Proteomics [1, 3, 20] and provide relevant information about genes and proteins. In short, all extracted data is represented by a matrix whose rows are pairs of proteins and whose columns are the calculated features.

Finally, we have also introduced similarity measures based on the common terms that two proteins share. As the results section will show, this idea will help us improve the power of the prediction in our approach.

3. Feature Selection

In this work, we have used two approaches to feature selection. In our first approach we made use of the Relief algorithm [31] in order to draw a model to evaluate our solution from several different datasets. This algorithm keeps a weight vector over all features and updates this vector according to the given sample points. Under some assumptions, Kira and Rendell demonstrated that the expected weight is large for relevant features and small for irrelevant ones. They also explained how to select a determinant threshold in a way that ensures a low probability that a given irrelevant feature is chosen.

We have chosen the implementation of Relief from Gilad-Bachrach et al. [32]. This algorithm presents several advantages: it is a well-known method, it has been used in Proteomics providing good results [33], it is simple and it is efficient. Given a dataset, Relief returns a ranking of features according to an importance weight. In order to obtain a subset including the most relevant features, we selected the features whose weight was above the mean weight of all features. This subset of features was then used for training and evaluating the performance of the resulting SVM models.

In the second approach, with the intention of testing more than one feature selection algorithm, we propose a new approach using a combination of well-known feature selection methods to select the relevant features and train SVM models. This novel method is based on 3 feature selection methods: G-flip, Simba [32] and the already mentioned Relief. G-flip and Simba assign a score to sets of features according to the induced margin from an evaluation function. This GFLIP and SIMBA evaluation function requires a utility function; we used three functions proposed by Navot et al. [32]: linear, sigmoid and zero-one. The linear utility function is a simple sum of margins. The Zero-one utility (not for SIMBA) equals to 1 when the margin is positive and 0 otherwise. And finally, the sigmoid utility function is proportional to the leave-one-out (LOO) error.

Our proposed feature selection method consists in normalizing the weights between [0,1] given by all previous methods (Gflip, Simba and Relief). Subsequently, the final weight for each feature is calculated as a mean of all weights of all methods. The feature with a final weight above the mean are the features selected to perform the classification (see figure 2(a)).

4. Support Vector Machines

Support Vector Machines (SVM) are a classification and regression paradigm first developed by Vapnik [34, 35]. The SVM approach is quite popular in the literature related to classification and regression problems because it provides a good generalization performance. In SVM, the target is to find an optimal sep-

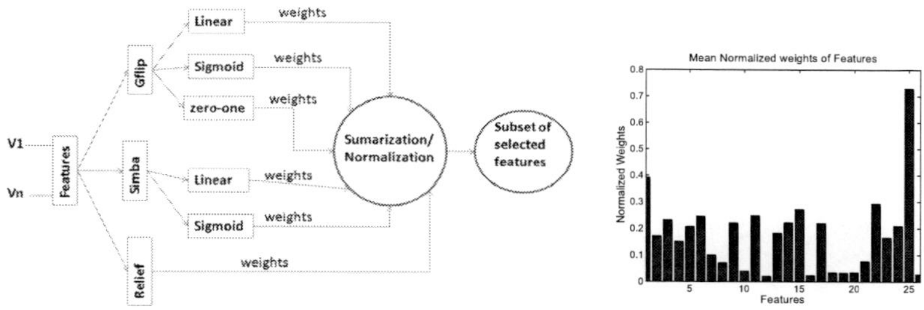

(a) (b)

Figure 2. a) Proposed feature selection. b) Feature weights.

arating hyperplane with low generalization error. Since SVMs were originally designed for binary-class classification, it is straightforward to use this paradigm in the present problem for classification among interacting and non-interacting pairs of proteins.

Given a training set of instance-label pairs $\{x_i, y_i\}$, $i = 1, ..., N$ with input data $x_i \in \mathbb{R}^n$ and labeled output data $y_i \in \{-1, +1\}$. The classification decision function for a linear SVM problem [36] is represented by Eq. 2:

$$y(x) = sign[\sum_{i=1}^{N} \alpha_i y_i x_i^T x + b] \tag{2}$$

where the α_i correspond to the solution of the Quadratic Programming (QP) problem that solves the maximum margin optimization problem. The training data points corresponding to non-zero α_i values are called *support vectors* [36].

For the nonlinear SVM classifier, in the decision function in equation Eq. 2, every dot product is replaced by a non-linear kernel (K) function. This allows the algorithm to fit a maximum-margin hyperplane in a transformed feature space; i.e., the classifier is a hyperplane in the high-dimensional feature space that may be non-linear in the original input space. The decision function in the nonlinear case [36] is represented by Eq. 3:

$$y(x) = sign[\sum_{i=1}^{N} \alpha_i y_i K(x, x_i) + b] \tag{3}$$

Again, the non-zero α_i values will correspond to the support vectors in the higher dimensional feature space.

Two typical kernels used in Bioinformatics, and that will be used in this work are: the Linear kernel and the Radial Basis Function (RBF) kernel [37]. The linear kernel is represented by:

$$K(x, x_i) = x_i^T x \tag{4}$$

that corresponds to the linear SVM. The Radial Basis Function (RBF) kernel is represented by:

$$K(x, x_i) = exp(-\gamma||x - x_i||^2), \gamma > 0 \tag{5}$$

Training an SVM implies the optimization of both the α_i, which can be optimally obtained by solving a QP problem, and of the so-called hyperparameters of the model, normally estimated using grid-search and Cross-validation [36]. For the Linear Kernel, only the hyperparameter $C > 0$ needs to be optimized, which is the penalty parameter of the error term, or in other words, it is a real positive constant that controls the amount of misclassification allowed in the model. For the RBF Kernel, both C and γ need to be optimized. For the simulations, we constructed SVM classifiers using the linear and the RBF Kernel, and we assessed the classification error for several high-quality datasets in order to compare their predicting power.

All the algorithms have been implemented in Matlab 2007a (R) using the library LIBSVM [38] for Support Vector Machines. Specifically, we have used C-SVM with linear and RBF kernels provided by this library.

5. Results

In this section, the proposed approach is evaluated using a number of specific high-quality datasets and check the two feature selection methods already presented in the previous sections.

For the first method, we can find two clearly different sub-parts in our experiment. In the first sub-part, to check the reliability of our SVM-based proposal, we have calculated specificity and sensitivity values for three training and testing partitions of the dataset presented in section 2.. For this case, we have obtained specificity and sensitivity values higher than 90% both for linear and RBF-based SVMs. In the second sub-part, we have constructed SVM predictors trained using the original full dataset extracted from Saeed et al. [14], and evaluated the classification accuracy for the binary interactions datasets analyzed in [23].

Following the steps explained in the section 2., 26 genomic/proteomic features were extracted from different databases, which included the additional features calculated by the proposed similarity measures. In order to obtain the most relevant features, we have applied the Relief algorithm (section 3.) and we have chosen the features whose returned weight was above the mean (as shown in figure 3). The final features were 9 out of the initial 26 attributes, accomplishing the goal of reducing the number of input variables and, as it will be shown, obtaining a low classification error. Table 1 shows the selected features which are: 25th referring to local similarity measure for MIPS phenotype catalog, 22nd referring to local similarity measure for MIPS functional catalog, 11th referring to similarity measure for sub-cellular compartments, 1st referring to number of GO common terms, 13th referring to local similarity measure for number of GO common terms, 23rd referring to local similarity measure for MIPS complexes catalog, 5th referring to number of GO common terms in the Biological Processes Ontology, 2nd referring to number of shared homologous proteins between a pair of proteins and 24th referring to local similarity measure for MIPS proteins catalog. Therefore, the whole data to consider is a 9704x9 matrix, whose rows are pairs of proteins and whose columns are the nine selected features.

Table 1. Features

Selected features by Relief
25, 22, 11, 1, 13, 23, 5, 2, 24

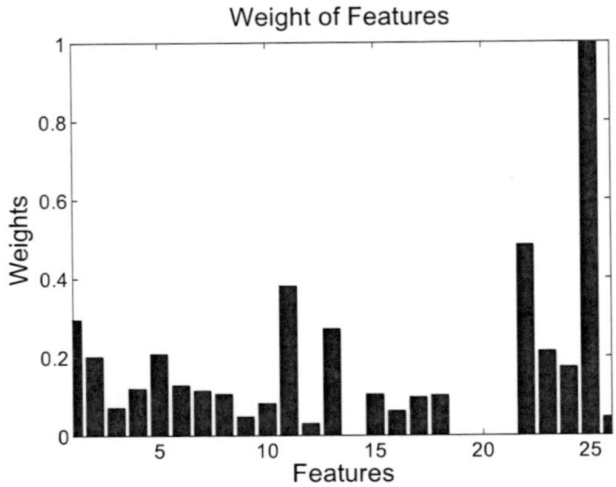

Figure 3. Normalized Weight of Features obtained by Relief

In order to evaluate the selected subset of features, we have randomly split the dataset extracted from Saeed et al. [14] into two groups: 70% of data for training and 30% for test. We have performed linear and RBF-based SVM classification for three executions with different randomly training and test partitions (as shown in table 2). A single combination of the calculated parameters using p-fold Cross-Validation (CV) with $p = 10$, γ and C parameters for the RBF Kernel and C for the linear kernel was employed for these executions.

Sensitivity is the capacity to properly classify an interacting pair and specificity is the capacity to properly classify a non-interacting pair; so in other words, the sensitivity measures the proportion of actual positives (interacting pairs of proteins) which are correctly identified. Specificity measures the proportion of negatives (non-interacting pairs) which are correctly identified. In the table 2, we can observe that SVM predictors obtain sensitivity values higher than 98% and specificity values higher than 90% for the different test sets.

Comparing the results of the two applied kernels in SVMs, the nonlinear SVM with RBF kernel returned the best results. RBF-based SVM was able to detect over 99% of protein-protein interactions, and the negative pairs over 92%. In this experiment our approach was more reliable in the classification of non-interacting pairs in comparison with other works in the literature (e.g. Patil et al. [20] obtained a 90% sensitivity and a 63% specificity using a Bayesian

Table 2. Classification Error with 3 Randomly Partitioned Datasets

Training/Test Group	Linear Kernel SVM		
	Error (%)	Se. (%)	Sp. (%)
1st	4.6	98.7	91.8
2nd	4.4	98.3	92.9
3rd	4.7	98.8	91.7
Mean	4.57	98.60	92.13
Std. Dev.	0.12	0.22	0.54

Training/Test Group	RBF Kernel SVM		
	Error (%)	Se. (%)	Sp. (%)
1st	4	99.1	92.7
2nd	3.6	99.2	93.5
3rd	4	99.5	92.5
Mean	3.8	99.27	92.90
Std. Dev.	0.19	0.17	0.43

Se: Sensitivity, Sp: Specificity. Error: Classification Error. Std Dev: Standard Deviation. Tr/Test: Training/Test.

approach). Linear SVM obtained slightly worse results but the execution was faster. RBF-based SVMs take longer both in the training and the test phases. So in case that performance is important, linear SVMs would be suitable because the results are quite similar to RBF-based SVMs.

In the second sub-part of our experiments, we checked the performance of our proposal using other high-quality datasets. To do so, we designed both a linear and an RBF-based SVM using the complete dataset described in section 2. as training set. As testing sets, we took the binary interactions datasets from Yu et al. [23] and calculated the classification error. In that work, a comparative assessment of current yeast interactome data sets was carried out using computational and experimental approaches. The authors demonstrated that High-Throughput Y2H screening provides high-quality binary interactions information. A framework was empirically developed to produce "a second-generation" high-quality dataset for yeast binary interactions. They also provided several binary interactions datasets: LC-multiple, Binary-GS, CCSB-Y2H, Y2H_union. For negative examples they also supplied the RRS dataset. These datasets can be downloaded from http://interactome.dfci.harvard.edu and their main characteristics are the following:

- The LC-multiple dataset is composed of only literature-curated interactions. Specifically, only those ones curated from two or more publications. They are 1253 positive interactions.
- Binary-GS dataset is a binary gold standard set that was assembled through a computational quality reexamination that includes well-established complexes, as well as conditional interactions and well-documented direct physical interactions in the yeast proteome. There are 2855 positive interactions.
- CCSB-Y2H dataset is the result of a new proteome-scale yeast high-throughput Y2H screen carried out by the authors. It is made up of 1725 positive interactions.
- Y2H_union dataset is a combination of three available high-quality proteome-scale Y2H datasets (up to 2815 positive interactions):
  - Uetz-screen: The union of sets found by Uetz et al. in a proteome-scale all-by-all screen [5].
  - Ito-core: interactions found by Ito et al. that appear three times or more [6].
  - CCSB-Y2H dataset commented before.
- RRS dataset (Yeast Random Reference Set). This set is composed of random paired proteins that the authors considered that are extremely unlikely to be interacting. They are 156 pairs of non-interacting proteins.

The obtained results in this second part of the experiments are shown in table 3. The RRS dataset is the only negative set, so the returned accuracy refers to the proportion of non-interacting pairs (negatives) which are correctly identified. The accuracy for only positive sets refers to the proportion of interacting pairs (positives) which are correctly identified.

We can observe that both linear SVM and RBF-based SVM achieved an accuracy higher than 87% in all datasets except the completely experimental dataset. As in the first part of this section, the linear SVM yielded slightly

Table 3. Accuracy in Prediction with Binary Interactions Datasets from Yu et al. [23]

Datasets	Linear Kernel SVM Accuracy (%)
LC-multiple	98.809
Binary-GS	98.244
CCSB-YI1	80.464
Y2H_union	87.815
RRS	100

Datasets	RBF Kernel SVM Accuracy (%)
LC-multiple	99.124
Binary-GS	98.404
CCSB-YI1	82.551
Y2H_union	89.094
RRS	100

lower accuracy values than RBF-based SVM. For the LC-multiple and Binary-GS datasets, this approach could predict them with around 99% of accuracy; Yu et al. [23] estimated the fraction of true positives of CCSB-Y2H around 94-100%, however we obtained only 80-82% of accuracy. However, when the information about the well-known datasets in the literature (Uetz et al. [5] and Ito et al. [6]) is incorporated to CCSB-Y2H forming the dataset Y2H_union, the accuracy obtained by our proposal grew 7% more, yielding an accuracy of almost 90% for the RBF-based SVM. Concerning the RRS negative set, our approach can detect the 100% of non-interacting pairs for both SVM classifiers, so we can say that our SVM approach is able to predict both interacting and non-interacting pairs of proteins.

In the second stage, we use our proposed feature selection method with the intention to demonstrate whether it leads to more accurate SVM models. Finally, we will assess the quality of the data giving a score for each pair of proteins and analyzing the results. Therefore following the steps of our experiment, we extracted all features from our initial dataset [14], and applied our feature selection method to obtain the subset of relevant features.

As previously, due to computational needs we use 70% of data randomly distributed for the feature selection methods. Simba and Gflip were executed over 10 iterations applying all their utility functions for the evaluation function. Finally, features that were selected all 10 times in Gflip were chosen. On the other hand, in Simba the features were selected according to mean weights in all executions. Relief is deterministic. Our proposed method calculated a mean weight per feature derived from the weights of these methods. Table 4 shows the selected features by method, and figure 2(b) shows the final feature weights for our proposed method.

Table 4. Table of selected features by different selection methods

Method	Selected Features
Gflip Linear	1 2 3 5 6 7 8 10 11 13 15 16 17 18 19 22 23 24 25
Gflip Sigmoid	1 2 3 4 5 6 7 8 9 10 11 13 15 16 17 18 20 21 22 23 24 25
Gflip zero-one	1 2 25
Relief	25 22 11 1 13 5 23 2 24
Simba Linear	3 15 22 25
Simba Sigmoid	1 3 4 6 11 15 22 23 25
Our Proposal	1 3 5 6 9 11 13 14 15 17 22 24 25

In the next step, the data was once more randomly divided into two groups: 70% pairs to train the SVMs, and 30% pairs to test the models. A single combination of calculated parameters γ and C for RBF Kernel and C for linear kernel was employed. The performance of the test set was evaluated by sensitivity and specificity. We executed three times our predictors with three different randomly distributed groups for training and test. We obtained for linear-SVMs: classification error 6,63 % $\pm$ 1,37, sensitivity 97,4 % $\pm$ 0,37 and specificity 89,17 % $\pm$ 2,5. For RBF-SVMs the results are: classification error 5,03 % $\pm$ 0,25, sensitivity 98,3 % $\pm$ 0,36, specificity 91,5 % $\pm$ 0,88.

In figure 4, we can observe the classification score of a test group for correct and wrong predicted pairs, determining clearly two groups. Correct predicted pairs obtain scores 0,905 $\pm$ 0,149, and as we expected the wrong predicted pairs obtain 0,584 $\pm$ 0,302. This demonstrates the reliability of our proposed SVM predictor, returning high scores for correctly classified interactions.

Table 5. Comparison table with results of Selected Features

LINEAR SVM	Our proposal with best features appeared in the bibliography [20]			Results by Patil [20]			Our Proposal with the 3 best features		
Features	Error (%)	Sv. (%)	Sp. (%)	Sv. (%)	Sp. (%)	Features	Error (%)	Sv (%)	Sp (%)
d+g+h	7,03±0,47	92,53±0,52	93,4±0,42	12,3	99,4	1+2+3	11,63±0,05	95,13±0,09	81,23±0,19
d+g	15,73±0,61	92,6±0,57	75,5±0,57	14,5	99,3	1+3	14±0,14	99,83±0,05	71,47±0,52
d+h	20,5±0,85	99,9±0	58±1,27	14,7	99,2	1+2	12,7±0	92,6±0,57	81,73±0,75
d	34,3±0,42	67,57±0,66	63,77±0,24	14,8	99,2	1	14,37±0,05	99,83±0,05	70,67±0,38
g+h	7,03±0,47	92,53±0,52	93,4±0,42	44,1	94	2+3	14,7±0,42	95,13±0,09	74,9±0,42
g	15,73±0,61	92,6±0,57	75,5±0,57	86,7	74,3	3	34,43±0,38	38,93±0,05	93,7±0
h	20,5±0,85	99,9±0	58±1,27	89,7	62,9	2	15,73± 0,61	92,6±0,57	75,5±0,57
RBFs SVM	Our proposal with best features appeared in the bibliography [20]			Results by Patil [20]			Our Proposal with the 3 best features		
Features	Error (%)	Sv. (%)	Sp. (%)	Sv. (%)	Sp. (%)	Features	Error (%)	Sv (%)	Sp (%)
d+g+h	7,03±0,47	92,53±0,52	93,4±0,42	12,3	99,4	1+2+3	7,53±0,24	98,9±0	85,67±0,66
d+g	15,37 ± 0,8	91,1±0,85	77,8±2,26	14,5	99,3	1+3	8,77±0,19	99,73±0,05	82,27±0,52
d+h	20,5±0,85	99,9±0	58±1,27	14,7	99,2	1+2	8,47±0,33	98,47±0,09	84,27±0,8
d	34,37±0,47	67,57±0,66	63,67±0,24	14,8	99,2	1	10,83±0,38	99,83±0,05	77,97±0,94
g+h	7,03±0,47	92,53±0,52	93,4±0,42	44,1	94	2+3	14±0,42	95,57±0,19	75,97±0,47
g	15,73±0,61	92,6±0,57	75,5±0,57	86,7	74,3	3	24,73±0,09	99,9±0	49,23±0,75
h	20,53±0,83	99,9±0	58±1,27	89,7	62,9	2	15,73±0,61	92,6±0,57	75,5±0,57

Sv: sensitivity, Sp: Specificity. 1: (25th feature) Phenotype MIPS calculated by our "local similarity" measure, 2:(1st *feature*) similar GO annotations, 3: (22th feature) Functional MIPS calculated by our "local similarity" measure .d: interacting Pfam domains (3rd feature); g: similar GO annotations (1st feature) ; h: homologous interactions (2st feature). More than one features are indicated by listing the features separated by a '+' sign.

To check that the selected features calculated by our similarity measure help to improve the predictive power of the SVM model, we performed a comparison between the three features proposed by Patil et al. [20] (d,g,h) and our three highest weighted features (25st, 1st, 22st) obtained with our approach (see figure 2(b)). Table 5 shows the results of this comparison.

As it can be observed in table 5, in general, our selection of three features does not improve the obtained results with the proposed features by Patil et al. [20]. However, we notice that more than 3 features are necessary to provide a correct classification, giving higher values of specificity and sensibility.

Though a ROC (Receiver Operating Characteristic) curve (see figure 5), using our score as a measure of the quality of the prediction, we represented the results of the RBF-SVM classifier using the 14 selected features by our feature selection approach (red) versus the best results of our RBF-SVM approach with the "d+g+h" features proposed by Patil et al. [20] (blue). We observe better levels of sensibility and specificity using our 14 features. Hence, using a reasonable subset of features from the 26 extracted characteristics, we improve the prediction than using only 3.

Two of the three most important features (25th, 22th) considered by our proposal are derived from our similarity measure for phenotype and functional MIPS catalog. We found that features related with domains have not an important weight in our selection and the features related to GO annotations do have an important weight. Anyway, we confirm that a combination of relevant features permit the construction of reliable classifiers. The designed SVM classifiers with the selected subset of features prove to be good predictors.

Finally, we should mention that both the linear kernel and the RBF kernel obtain very good predictions in the test set. However as we expected, the RBF kernel is slightly more reliable.

In summary, our proposal provides reliable predictors of PPIs. The constructed classifiers are able to detect interacting and non-interacting pairs for the tested datasets obtained through well-known literature and computational approaches. The accuracy of SVMs in this experiments reached values higher than 90% except for the CCSB-Y2H. We also gave a "confidence score" that permits the validation and analysis of data.

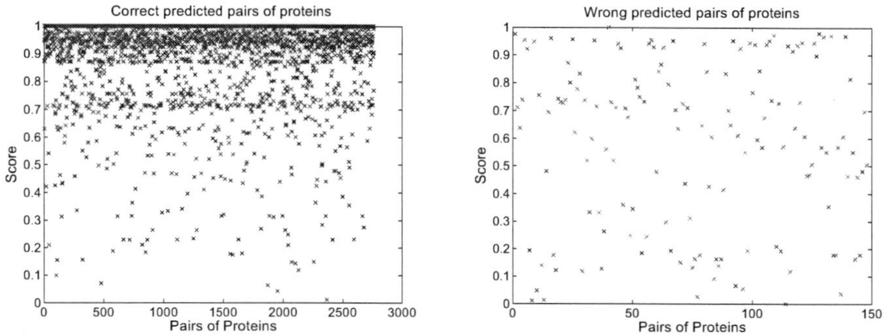

Figure 4. Scores for correct predicted pairs (left) and not correct predicted pairs (right)

6. Conclusions

In this work, we have proposed the use of features based on new similarity measures extracted from known databases: SwissPfam, GOA, MIPS, 3did, Hintdb. We have also proposed a new method of feature selection using a combination of classic feature selection algorithms. In the approach, we have used a set of examples extracted from a reliable source [14] for the organism of yeast (S. cerevisiae). Through our proposed feature selection method, we have chosen the relevant features in order to construct SVM classifiers as effective predictors, which may provide low error rate in classification of PPIs. We have also checked that the features derived from our similarity measures improved the predictive power of the classifiers. We have obtained high specificity and sensitivity in predicting PPIs from the high-quality dataset [14]. The quality of our SVM approach was evaluated using yeast binary interactions datasets from Yu et al. [23] obtaining reliable predictions for those curated-literature datasets. Finally, we have also provided a "confidence score" for each pair of proteins in the classification. Although this measure or score does not have to exclude other methodologies for validation, it could help to support results or avoid tedious and arduous validations in the laboratory, and hence, save money and time.

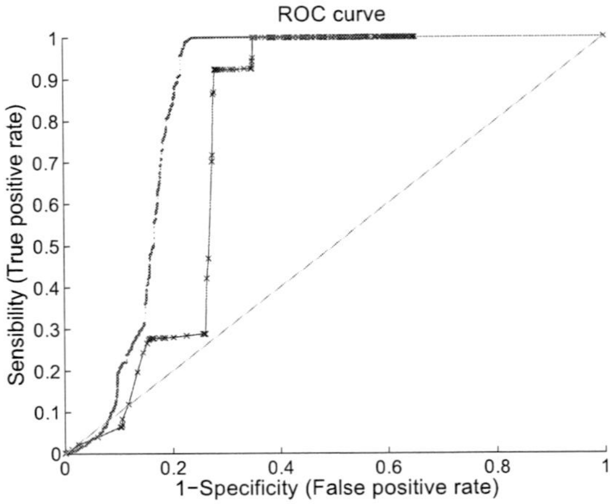

Figure 5. ROC curves. Our proposal (red) and a RBF-SVM with features appeared in Patil et al. [20] (blue)

Acknowledgments

José Miguel Urquiza Ortiz is supported by the FPU research grant AP2006-01748 from the Spanish Ministry of Education. This work has been partially supported by the Spanish CICYT Project TIN2007-60587 and Junta de Andalucía (Spanish Regional Government in Andalusia) Project P07-TIC-02768.

References

[1] C. Huang, F. Morcos, S. P. Kanaan, S. Wuchty, D. Z. Chen, and J. A. Izaguirre, "Predicting protein-protein interactions from protein domains using a set cover approach," *IEEE/ACM Transactions on Computational Biology and Bioinformatics / IEEE, ACM*, vol. 4, no. 1, pp. 78–87, 2007, PMID: 17277415.

[2] U. Stelzl and E. E. Wanker, "The value of high quality protein-protein interaction networks for systems biology," *Current Opinion in Chemical Biology*, vol. 10, no. 6, pp. 551–558, Dec. 2006, PMID: 17055769.

[3] R. Jansen, H. Yu, D. Greenbaum, Y. Kluger, N. J. Krogan, S. Chung, A. Emili, M. Snyder, J. F. Greenblatt, and M. Gerstein, "A bayesian networks approach for predicting Protein-Protein interactions from genomic data," *Science*, vol. 302, no. 5644, pp. 449–453, Oct. 2003.

[4] P. Braun, M. Tasan, M. Dreze, M. Barrios-Rodiles, I. Lemmens, H. Yu, J. M. Sahalie, R. R. Murray, L. Roncari, A. de Smet, K. Venkatesan, J. Rual, J. Vandenhaute, M. E. Cusick, T. Pawson, D. E. Hill, J. Tavernier, J. L. Wrana, F. P. Roth, and M. Vidal, "An experimentally derived confidence score for binary protein-protein interactions," *Nat Meth*, vol. 6, no. 1, pp. 91–97, 2009.

[5] P. Uetz, L. Giot, G. Cagney, T. A. Mansfield, R. S. Judson, J. R. Knight, D. Lockshon, V. Narayan, M. Srinivasan, P. Pochart, A. Qureshi-Emili, Y. Li, B. Godwin, D. Conover, T. Kalbfleisch, G. Vijayadamodar, M. Yang, M. Johnston, S. Fields, and J. M. Rothberg, "A comprehensive analysis of protein-protein interactions in saccharomyces cerevisiae," *Nature*, vol. 403, no. 6770, pp. 623–627, Feb. 2000.

[6] T. Ito, T. Chiba, R. Ozawa, M. Yoshida, M. Hattori, and Y. Sakaki, "A comprehensive two-hybrid analysis to explore the yeast protein interactome," *Proceedings of the National Academy of Sciences of the United States of America*, vol. 98, no. 8, pp. 4569–4574, Apr. 2001.

[7] A. Gavin, M. Bosche, R. Krause, P. Grandi, M. Marzioch, A. Bauer, J. Schultz, J. M. Rick, A. Michon, C. Cruciat, M. Remor, C. Hofert, M. Schelder, M. Brajenovic, H. Ruffner, A. Merino, K. Klein, M. Hudak, D. Dickson, T. Rudi, V. Gnau, A. Bauch, S. Bastuck, B. Huhse, C. Leutwein, M. Heurtier, R. R. Copley, A. Edelmann, E. Querfurth, V. Rybin, G. Drewes, M. Raida, T. Bouwmeester, P. Bork, B. Seraphin, B. Kuster, G. Neubauer, and G. Superti-Furga, "Functional organization of the yeast proteome by systematic analysis of protein complexes," *Nature*, vol. 415, no. 6868, pp. 141–147, 2002.

[8] Y. Ho, A. Gruhler, A. Heilbut, G. D. Bader, L. Moore, S. Adams, A. Millar, P. Taylor, K. Bennett, K. Boutilier, L. Yang, C. Wolting, I. Donaldson, S. Schandorff, J. Shewnarane, M. Vo, J. Taggart, M. Goudreault,

B. Muskat, C. Alfarano, D. Dewar, Z. Lin, K. Michalickova, A. R. Willems, H. Sassi, P. A. Nielsen, K. J. Rasmussen, J. R. Andersen, L. E. Johansen, L. H. Hansen, H. Jespersen, A. Podtelejnikov, E. Nielsen, J. Crawford, V. Poulsen, B. D. Sorensen, J. Matthiesen, R. C. Hendrickson, F. Gleeson, T. Pawson, M. F. Moran, D. Durocher, M. Mann, C. W. V. Hogue, D. Figeys, and M. Tyers, "Systematic identification of protein complexes in saccharomyces cerevisiae by mass spectrometry," *Nature*, vol. 415, no. 6868, pp. 180–183, 2002.

[9] S. Li, C. M. Armstrong, N. Bertin, H. Ge, S. Milstein, M. Boxem, P. Vidalain, J. J. Han, A. Chesneau, T. Hao, D. S. Goldberg, N. Li, M. Martinez, J. Rual, P. Lamesch, L. Xu, M. Tewari, S. L. Wong, L. V. Zhang, G. F. Berriz, L. Jacotot, P. Vaglio, J. Reboul, T. Hirozane-Kishikawa, Q. Li, H. W. Gabel, A. Elewa, B. Baumgartner, D. J. Rose, H. Yu, S. Bosak, R. Sequerra, A. Fraser, S. E. Mango, W. M. Saxton, S. Strome, S. van den Heuvel, F. Piano, J. Vandenhaute, C. Sardet, M. Gerstein, L. Doucette-Stamm, K. C. Gunsalus, J. W. Harper, M. E. Cusick, F. P. Roth, D. E. Hill, and M. Vidal, "A map of the interactome network of the metazoan c. elegans," *Science*, vol. 303, no. 5657, pp. 540–543, 2004.

[10] A. J. M. Walhout, R. Sordella, X. Lu, J. L. Hartley, G. F. Temple, M. A. Brasch, N. Thierry-Mieg, and M. Vidal, "Protein interaction mapping in c. elegans using proteins involved in vulval development," *Science*, vol. 287, no. 5450, pp. 116–122, 2000.

[11] L. Giot, J. S. Bader, C. Brouwer, A. Chaudhuri, B. Kuang, Y. Li, Y. L. Hao, C. E. Ooi, B. Godwin, E. Vitols, G. Vijayadamodar, P. Pochart, H. Machineni, M. Welsh, Y. Kong, B. Zerhusen, R. Malcolm, Z. Varrone, A. Collis, M. Minto, S. Burgess, L. McDaniel, E. Stimpson, F. Spriggs, J. Williams, K. Neurath, N. Ioime, M. Agee, E. Voss, K. Furtak, R. Renzulli, N. Aanensen, S. Carrolla, E. Bickelhaupt, Y. Lazovatsky, A. DaSilva, J. Zhong, C. A. Stanyon, R. L. Finley, K. P. White, M. Braverman, T. Jarvie, S. Gold, M. Leach, J. Knight, R. A. Shimkets, M. P. McKenna, J. Chant, and J. M. Rothberg, "A protein interaction map of drosophila melanogaster," *Science*, vol. 302, no. 5651, pp. 1727–1736, Dec. 2003.

[12] E. Formstecher, S. Aresta, V. Collura, A. Hamburger, A. Meil, A. Trehin, C. Reverdy, V. Betin, S. Maire, C. Brun, B. Jacq, M. Arpin, Y. Bellaiche, S. Bellusci, P. Benaroch, M. Bornens, R. Chanet, P. Chavrier, O. Delattre, V. Doye, R. Fehon, G. Faye, T. Galli, J. Girault, B. Goud, J. de Gunzburg, L. Johannes, M. Junier, V. Mirouse, A. Mukherjee, D. Papadopoulo, F. Perez, A. Plessis, C. Rosse, S. Saule, D. Stoppa-Lyonnet, A. Vincent, M. White, P. Legrain, J. Wojcik, J. Camonis, and L. Daviet, "Protein interaction mapping: A drosophila case study," *Genome Research*, vol. 15, no. 3, pp. 376–384, Mar. 2005, PMC551564.

[13] T. Bouwmeester, A. Bauch, H. Ruffner, P. Angrand, G. Bergamini, K. Croughton, C. Cruciat, D. Eberhard, J. Gagneur, S. Ghidelli, C. Hopf, B. Huhse, R. Mangano, A. Michon, M. Schirle, J. Schlegl, M. Schwab, M. A. Stein, A. Bauer, G. Casari, G. Drewes, A. Gavin, D. B. Jackson, G. Joberty, G. Neubauer, J. Rick, B. Kuster, and G. Superti-Furga, "A physical and functional map of the human TNF-[alpha]/NF-[kappa]B signal transduction pathway," *Nat Cell Biol*, vol. 6, no. 2, pp. 97–105, Feb. 2004.

[14] R. Saeed and C. Deane, "An assessment of the uses of homologous interactions," *Bioinformatics*, vol. 24, no. 5, pp. 689–695, Mar. 2008.

[15] P. Bork, L. J. Jensen, C. von Mering, A. K. Ramani, I. Lee, and E. M. Marcotte, "Protein interaction networks from yeast to human," *Current Opinion in Structural Biology*, vol. 14, no. 3, pp. 292–299, Jun. 2004.

[16] C. M. Deane, L. Salwinski, I. Xenarios, and D. Eisenberg, "Protein interactions: Two methods for assessment of the reliability of high throughput observations," *Mol Cell Proteomics*, vol. 1, no. 5, pp. 349–356, May 2002.

[17] C. von Mering, R. Krause, B. Snel, M. Cornell, S. G. Oliver, S. Fields, and P. Bork, "Comparative assessment of large-scale data sets of protein-protein interactions," *Nature*, vol. 417, no. 6887, pp. 399–403, May 2002.

[18] J. S. Bader, A. Chaudhuri, J. M. Rothberg, and J. Chant, "Gaining confidence in high-throughput protein interaction networks," *Nat Biotech*, vol. 22, no. 1, pp. 78–85, 2004.

[19] J. Yu and R. L. Finley, "Combining multiple positive training sets to generate confidence scores for protein-protein interactions," *Bioinformatics*, vol. 25, no. 1, pp. 105–111, 2009.

[20] A. Patil and H. Nakamura, "Filtering high-throughput protein-protein interaction data using a combination of genomic features," *BMC Bioinformatics*, vol. 6, no. 1, p. 100, 2005.

[21] X. Wu, L. Zhu, J. Guo, D. Zhang, and K. Lin, "Prediction of yeast protein-protein interaction network: insights from the gene ontology and annotations," *Nucl. Acids Res.*, vol. 34, no. 7, pp. 2137–2150, Apr. 2006.

[22] G. O. Consortium, "The gene ontology (GO) database and informatics resource," *Nucl. Acids Res.*, vol. 32, no. suppl_1, pp. D258–261, 2004.

[23] H. Yu, P. Braun, M. A. Yildirim, I. Lemmens, K. Venkatesan, J. Sahalie, T. Hirozane-Kishikawa, F. Gebreab, N. Li, N. Simonis, T. Hao, J. Rual, A. Dricot, A. Vazquez, R. R. Murray, C. Simon, L. Tardivo, S. Tam, N. Svrzikapa, C. Fan, A. de Smet, A. Motyl, M. E. Hudson, J. Park, X. Xin, M. E. Cusick, T. Moore, C. Boone, M. Snyder, F. P. Roth, A. Barabasi, J. Tavernier, D. E. Hill, and M. Vidal, "High-Quality binary protein interaction map of the yeast interactome network," *Science*, vol. 322, no. 5898, pp. 104–110, Oct. 2008.

[24] B. Boeckmann, A. Bairoch, R. Apweiler, M. Blatter, A. Estreicher, E. Gasteiger, M. J. Martin, K. Michoud, C. O'Donovan, I. Phan, S. Pilbout, and M. Schneider, "The SWISS-PROT protein knowledgebase and its supplement TrEMBL in 2003," *Nucl. Acids Res.*, vol. 31, no. 1, pp. 365–370, 2003.

[25] E. Camon, M. Magrane, D. Barrell, V. Lee, E. Dimmer, J. Maslen, D. Binns, N. Harte, R. Lopez, and R. Apweiler, "The gene ontology annotation (GOA) database: sharing knowledge in uniprot with gene ontology," *Nucleic Acids Research*, vol. 32, no. Database issue, pp. D262–D266, 2004, PMC308756.

[26] U. Guldener, M. Munsterkotter, G. Kastenmuller, N. Strack, J. van Helden, C. Lemer, J. Richelles, S. J. Wodak, J. Garcia-Martinez, J. E. Perez-Ortin,

H. Michael, A. Kaps, E. Talla, B. Dujon, B. Andre, J. L. Souciet, J. D. Montigny, E. Bon, C. Gaillardin, and H. W. Mewes, "CYGD: the comprehensive yeast genome database," *Nucl. Acids Res.*, vol. 33, no. suppl_1, pp. D364–368, 2005.

[27] A. Patil and H. Nakamura, "HINT - a database of annotated protein-protein interactions and their homologs," *Biophysics*, 2005.

[28] A. Stein, R. B. Russell, and P. Aloy, "3did: interacting protein domains of known three-dimensional structure," *Nucl. Acids Res.*, vol. 33, no. suppl_1, pp. D413–417, 2005.

[29] H. M. Berman, J. Westbrook, Z. Feng, G. Gilliland, T. N. Bhat, H. Weissig, I. N. Shindyalov, and P. E. Bourne, "The protein data bank," *Nucl. Acids Res.*, vol. 28, no. 1, pp. 235–242, 2000.

[30] Y. Liu, I. Kim, and H. Zhao, "Protein interaction predictions from diverse sources," *Drug Discovery Today*, vol. 13, no. 9-10, pp. 409–416, May 2008.

[31] K. Kira and L. A. Rendell, "A practical approach to feature selection," in *Proceedings of the ninth international workshop on Machine learning* . Aberdeen, Scotland, United Kingdom: Morgan Kaufmann Publishers Inc., 1992, pp. 249–256.

[32] A. N. R. Gilad-Bachrach and N. Tishby, "Margin based feature selection: Theory and algorithms," *In Proc. of the 21'st ICML*, pp. 43–50, 2004.

[33] P. Block, J. Paern, E. Hullermeier, P. Sanschagrin, C. A. Sotriffer, and G. Klebe, "Physicochemical descriptors to discriminate protein-protein interactions in permanent and transient complexes selected by means of machine learning algorithms," *Proteins: Structure, Function, and Bioinformatics*, vol. 65, no. 3, pp. 607–622, 2006.

[34] C. Cortes and V. Vapnik, "Support vector network," *Mach. Learn.*, 1995.

[35] L. Herrera, H. Pomares, I. Rojas, A. Guillen, A. Prieto, and O. Valenzuela, "Recursive prediction for long term time series forecasting using advanced models," *Neurocomputing*, vol. 70, no. 16-18, pp. 2870–2880, Oct. 2007.

[36] J. A. K. Suykens, T. V. Gestel, J. D. Brabanter, B. D. Moor, and J. Vandewalle, *Least Squares Support Vector Machines*. World Scientific Publishing Company, 2003.

[37] I. Rojas, H. Pomares, J. Gonzales, J. L. Bernier, E. Ros, F. J. Pelayo, and A. Prieto, "Analysis of the functional block involved in the design of radial basis function networks," *Neural Processing Letters*, vol. 12, no. 1, pp. 1–17, 2000.

[38] C. Chang and C. Lin, *LIBSVM: a Library for Support Vector Machines*, 2001, software available at http://www.csie.ntu.edu.tw/ cjlin/libsvm.

In: Proteomics
Editor: Giselle C. Rancourt
ISBN: 978-1-61668-691-8

Chapter 6

Label-Free Liquid Chromatography-Based Quantitative Proteomics: Challenges and Recent Developments

A. Matros*[*1], S. Kaspar[1], S. Tenzer[2], M. Kipping[3], U. Seiffert[4] and H.-P. Mock[1]

[1]Leibniz Institute of Plant Genetics and Crop Plant Research, Gatersleben, Germany
[2]Institute of Immunology, University of Mainz, Mainz, Germany
[3]Waters GmbH, Eschborn, Germany
[4]Fraunhofer-Institute IFF Magdeburg, Biosystems Engineering, Magdeburg, Germany

Abstract

Recent innovations in liquid chromatography-mass spectrometry (LC-MS) based methods have facilitated comparative and functional proteomic analyses of large numbers of proteins derived from complex

---

* Corresponding Author: Andrea Matros, Leibniz Institute of Plant Genetics and Crop Plant Research, Dept. Molecular and Cell Physiology, Applied Biochemistry Group, Corrensstrasse 3, D-06466, Gatersleben, Germany, Phone: 49 39482 5-445, Fax:+49 39482 5-524

samples without any need for protein or peptide labelling. Here we discuss the features of label-free LC-based proteomics techniques. We first summarize recent methods used for quantitative protein analyses by MS techniques. The major challenges faced by label-free LC-MS based approaches are discussed; these include sample preparation, peptide separation, data mining and quantification. Absolute quantification, kinetic approaches and database search algorithms are also addressed. We focus on the Expression^{E} System^{TM} (Waters, Manchester, UK), a relatively new platform allowing label-free quantification of peptides for which mass and retention time have been accurately measured. Enhancing the power of this method will require developments in both separation technology and bioinformatics/statistical analysis.

Recent Methods Used in Quantitative Proteomics

Differential quantitative proteomics can deliver a clearer picture of molecular biological processes than is possible using a global transcriptomic approach. Recent years have seen the development of a number of novel ways of accessing the complete proteome expressed in particular tissues, cells or sub-cellular fractions (Table 1); these generally complement rather than replace the well-established 2-D gel electrophoresis approach (Lilley and Dupree 2006; Thelen and Peck 2007; Bachi and Bonaldi 2008; Mann 2009; Oeljeklaus et al. 2009; Wilm 2009). The identification of individual proteins, protein complexes, and the protein composition of specific organelles and tissues is now at the cutting edge of this area of technology. A start has been made to characterize the entire cellular proteome of certain organisms, and even to profile protein-protein interactions (Gingras et al. 2005; Selbach and Mann 2006; de Godoy et al. 2008; Sharma et al. 2009; Wessels et al. 2009). These early forays into the proteome have shown that the situation is rather more complex, diverse and dynamic than had been originally anticipated. As a result, it has become common practice to apply mass spectometry (MS) based proteomic approaches to complex mixtures as well as to pre-fractionated extracts in order to identify post-translationally modified peptides (Johnson and Hunter 2004; Jensen 2006; Bodenmiller et al. 2007; Oeljeklaus et al. 2009). The complete description of a proteome may remain forever out of reach, given the level of complexity introduced by the presence of mRNA splicing, various post-translational modifications and variable degradation products (Picotti et al. 2007). A recent analytical development has been multiple reaction monitoring (MRM) mass spectrometry, which seeks to validate global proteomic data, and to characterize post-translational modifications (Hewel and Emili 2008; Addona et al. 2009; Yocum and Chinnaiyan 2009). It has also been applied in certain large- scale targeted analyses of full proteomes (Picotti et al. 2009).

The quantification of both the relative and absolute amounts of particular proteins/peptides is a critical goal of any proteomic technique. The use of MS data for this purpose is not straightforward, and a number of strategies have therefore been elaborated to address this issue (Bachi and Bonaldi 2008; Wilm 2009). The signals used for quantification are either derived from the intact peptide (MS) or from one or more of its fragments (MS/MS). The latter suffers from a lesser level of interference from background ions, but signal intensity is often too low to allow sufficient precision (Wilm 2009). Thus, most quantitative MS applications tend to rely on MS, rather than on MS/MS. The quantification of MS signals can be based either on using isotopic reference peptides, or global references. In the former case, the measured value is compared to that of an isotopically labelled peptide of similar molecular structure. Typically, samples to be compared are labelled differentially, and then combined prior to analysis. This labelling can be achieved either in vivo, or by in vitro treatment of cell extracts. When global references are used for quantification, the measured value of each peptide is related to a set of molecules which are chemically distinct from the target (Table 1). This method is referred to as label-free, or direct quantification (Wang, W. et al. 2003; Silva et al. 2005). As this approach is limited to electrospray applications, it is dependent on stable, reproducible and accurate chromatography platforms. With the development of a range of instruments dedicated to electrospray ionization (ESI) MS of proteins and peptides, and the introduction of improved, miniaturized liquid chromatography (nano-LC) systems, proteomic approaches based on LC-ESI MS have become increasingly popular (Bachi and Bonaldi 2008; Levin et al. 2009).

Spectral counting is a correlation-based means of determining relative protein quantity (Liu et al. 2004). The technique is based on the observed correlation between the amount of a given peptide present in a sample and the frequency with which it can be fragmented in an ion trap MS. The approach is applied in parallel with a quantification method based on the total ion intensity of the target peptide (Old et al. 2005; Fang et al. 2006; Zhang et al. 2006; Wilm 2009).

A recently introduced alternative to label-free quantification approaches using ion trap MS data has been the use of a specific MS acquisition mode based on parallel fragmentation (Silva et al. 2005; Huges et al. 2006; Cutillas and Vanhaesebroeck 2007; Gilar, M. et al. 2009). Here, data acquisition is conducted by running two alternating LC-MS traces which differ from one another with respect to their applied collision energy (low *vs.* high collision energy, MS^E). The low energy mode provides accurate mass and quantitative data at the peptide level, while the high energy mode provides MS fragmentation data of co-eluting peptides. The identification of the equivalent peptide relies both on the molecular ion (from the low collision energy trace) and fragment spectrum (from the elevated collision energy trace) information. Apart from the major advantage of acquiring data in terms of the duty cycle and the quantification possible, the initial specificity of these multiplexed LC-MS runs is less than that associated with a data dependent LC-MS/MS experiment. Nevertheless, specificity can be largely recovered by exploiting the elution profiles, the quantification of precursors and fragment ions, and various physicochemical properties evaluated by the application of a specialized ion accounting algorithm during peptide and protein identification (Li et al. 2009). A good level of compatibility between the outcomes of a data independent multiplexed LC-MS acquisition experiment and those of a data dependent LC-MS/MS (DDA) analysis has been shown recently (Geromanos et al. 2009). Such a comparison was based on a simple four protein mixture in the presence and absence of a complex tryptic digest from *Escherichia coli.* These samples were analysed in triplicate by both LC-MS/MS (DDA) and multiplexed LC-MS. Each individual set of data-independent LC-MS data identified a more comprehensive set of detected ions than the combined peptide identification derived from the DDA LC-MS/MS experiments. In the presence of the *E. coli* contaminant, >90% of the monoisotopic masses from the combined LC-MS/MS identifications were associated with their expected retention time. In addition, the fragmentation pattern and number of associated elevated energy product ions in each replicate experiment was very similar to the DDA identifications (Geromanos et al. 2009).

The resulting data are subsequently processed by taking into account ion detection, mass retention time pair clustering and normalization (Silva et al. 2006). The quantification procedure applied in the multiplexed LC-MS acquisition strategy can be performed at either the peptide or the protein level. Quantification at the protein level requires the association of each detected peptide with its related protein. The pooled peptides derived from a given protein are then used to quantify the protein. For quantification at the peptide level, each component is matched by accurate mass and retention time signatures.

Table 1. A comparison of current quantitative proteomics methods, as modified by (Wilm 2009)

	Method	Quantification based on	Capacity	Advantages	Limitations
Gel-based	2-DE	Protein spot intensities	1000-2000 protein spots	-Easy mass spectrometric protein identification -Unlimited number of samples to compare	-Low sensitivity -Gel-to-gel variance -Underrepresentation of extreme proteins (high mass, extreme pH, hydrophobic membrane proteins)
	DIGE	Protein spot intensities	1000-2000 protein spots	-High sensitivity -No gel variability	-Limited number of samples to compare (only three different dyes) -Slight mass shifts below 20 kDa -Challenging protein identification -Underrepresentation of extreme proteins
MS-based (label-free)	Peak-integration	Peptide intensities	500 proteins in 1D LC-MS	-Accuracy over high dynamic range -No general limitations in number of multiplexes -Low costs per sample -No chemical manipulation of samples -Can be used as absolute quantification method	-Reproducible sample preparation needed -Limited to ESI applications
	Spectral-counts	Number of MS/MS scans per precursor ion	500 proteins in 1D LC-MS	-Easy to achieve -Low costs per sample -No chemical manipulation of samples -Can be used as absolute quantification method	-High variance for low abundant signals -Semi-quantitative method -Reproducible sample preparation needed -Limited to ESI applications
MS-based (labeled)	^{15}N*	Peptide intensities	500 proteins in 1D LC-MS	-Reduced technical variation during fractionation and separation -High sensitivity -Quantifies *in vivo* changes	-Only two or three samples to be compared at the same time -Full metabolic labelling is difficult to achieve -Increased complexity in the MS-scan -High costs for large scale experiments
	SILAC*	Peptide intensities	500 proteins in 1D LC-MS	-Reduced technical variation during fractionation and separation -Quantifies *in vivo* changes -High sensitivity -can be used as absolute quantification method	-Only two or three samples to be compared at the same time -Full metabolic labelling is difficult to achieve -Increased complexity in the MS-scan -Applicable only to cell cultures -High costs for large scale experiments

Table 1. (Continued)

	Method	Quantification based on	Capacity	Advantages	Limitations
	^{18}O -trypsin	Peptide intensities	500 proteins in 1D LC-MS	-Reduced technical variation during fractionation and separation -Low costs -Always complete labeling reaction -High sensitivity	-Only pair wise comparison -Increased complexity in the MS-scan -Double incorporation of ^{18}O can occur
	ICAT	Peptide intensities	500 proteins in 1D LC-MS	-Reduced technical variation during fractionation and separation -High sensitivity	-Only Cys-containing proteins, -Influenced by cystein oxidation and beta-elimination -Increased complexity in the MS-scan -High costs
	iTRAQ	Fragment intensities	500 proteins in 1D LC-MS	-Reduced technical variation during fractionation and separation -Kinetic analysis possible (8-plex experiments) -High sensitivity -Nearly complete labelling	-Increased complexity in the MS-scan -Only fragmented peptides can be quantified -High costs
	AQUA	Peptide intensities	depends on number of available standards	-Reduced technical variation during fractionation and separation -High sensitivity -Absolute quantification method	-Increased complexity in the MS-scan -No compensation for sample losses -Quantification based on only one or two peptides per protein -High costs
	ICPL	Peptide intensities	500 proteins in 1D LC-MS	-Nearly complete labelling -High sensitivity -Basically no loss of sample	-Relatively large peptides after tryptic digestion because all lysine residues are blocked -Number of labels per peptide is sequence dependent -High costs

*in vitro

The peptides are clustered across all LC-runs and between samples, and finally the intensity of the peptide signals (de-isotoped and charge state-reduced) are used for quantification.

Comparative proteomic analyses tended to be limited to approaches requiring labelling, such as SILAC, ICAT and AQUA, in the past. However, an increasing number of experiments is now being attempted using a label-free approach. We have used a system developed by the Waters Corporation, which combines multiplexed LC-MS with label-free quantification, and is implemented into the ProteinLynx Global Server software (PLGS2.4, Waters, Manchester, UK). Our (and others') experience with this platform, and specific aspects of this approach are discussed in the following sections.

SAMPLE PREPARATION

Sample preparation from peptide mixtures is a critical step for full advantage to be taken of recent advances in high resolution LC-MS. Substances interfering with reproducible separation and/or MS detection need to be heavily diluted if not completely removed, since their presence can suppress the signal obtained from the target peptide(s), and thereby reduce the level of achievable sensitivity and reproducibility. The requirement for highly purified samples arises from the application of nanoscale chromatographic separation techniques to proteome studies. Particle-free sample preparation is essential, especially when on line desalting on a pre-column is not used. However, the additional steps introduced to ensure sample purity can often result in a significant degree of sample loss (Wang, H. et al. 2005). Therefore protocols especially for the preparation of small sample amounts, like micro-dissected material, have to be chosen carefully. Among the various alternative approaches described to date, the introduction of filtration devices and self-packed nano-scale columns appear to be the most promising. Filtration devices provide the extra benefit of being able to remove detergent from the sample solute, to permit the exchanging of buffer, and to enable protein digestion (Manza et al. 2005; Wisniewski et al. 2009). We have used in-filter digestion to process complex protein mixtures extracted from seeds, leaves, roots and cell suspension cultures, all of which typically contain appreciable concentrations of salts and carbohydrates such as sucrose and starch. In-filter digestion is also appropriate for samples containing nucleic acids (such as nuclear or plastid extracts). Self-packed nano-LC columns were pioneered by

(Gobom et al. 1999) for the purification of peptide extracts prior to MALDI-TOF MS. The extension of this methodology to samples prepared for LC-based separation has been described more recently (Rappsilber et al. 2007). In addition to its speed and robustness, its prime advantage lies in its flexibility in the choice of column resin (C4, C8, C18, SCX, etc.), which can extend the potential of the method by allowing for serial pre-fractionation.

When LC-MS is applied to protein analysis, detergents are required to ensure full solubilization and unfolding prior to digestion and these detergents must not interfere with the subsequent ionization process or MS analysis. Detergents commonly used to extract hydrophobic proteins must be removed before MS (Yeung et al. 2008). A number of effective MS compatible protein solubilizers (Invitrosol™, Invitrogen; RapiGest™, Waters, Protease MAX™, Promega, among others) have, however, in the meantime been marketed, and these are claimed to not require removal or to be easy to remove prior to MS analysis. In our hands, 0.5 % (w/v) RapiGest has been found to be an effective solubilizer of hydrophobic proteins.

Enhanced Data Quality by Improved LC-Separation

One of the major challenges facing protein profiling is the separation of highly complex peptide mixtures by LC prior to analysis by MS. The field of quantitative proteomics was opened up by the development and commercial availability of capillary columns suitable for reversed phase (RP) separation of tryptic peptides (Cooper et al. 2004; Kirkland and De Stefano 2006; Wang, X. et al. 2006). To deal with the complexity of proteomic samples, meanwhile, new ways have been discovered to enhance chromatography efficiency, and in particular, a reduction in the particle size and an increase in the column length. Well established parameters (e.g., increasing the column length and/or solvent temperature, and varying the gradient) are also effective. However, the inclusion of a sub 2μm particle stationary phase produces a large increase in back pressure, as optimal linear velocity is inversely proportional to particle size, and the column back pressure is inversely proportional to the square of particle size – thus, for example, a change from a 3.5 to a 1.7μm stationary phase particle size increases the back pressure eight fold. Thus, the development of ultra-performance LC (UPLC) systems was a critical factor in the dramatic increase in chromatographic performance achieved over the last decade. Increased performance produces an improved peak capacity, an increased speed and sensitivity, and enhances spectral quality due to the associated reduction in ionization suppression, as demonstrated recently (Wilson et al. 2005). In proteomics experiments, the platform is especially suited to the analysis of hydrophobic peptides (Scheurer et al. 2005; Krämer-Albers et al. 2007; Jahn et al. 2009) and peptide isomers (Winter et al. 2009).

To take advantage of the enhanced levels of detection allowed by current analytical techniques, both peptide separation (Figure 1) and database searching (see below under "Databases and Search Algorithms" and Figure 3) needed to be optimized. Thus, we set out to improve the chromatographic resolution of highly complex mixtures of tryptic peptides, aiming to reduce ion suppression effects during the electrospray process and to increase the information yield from each sample. We applied an LC-based approach, combined with ESI-Q-TOF MS. A comparison was made between the separation effectiveness from a mixture of tryptic barley grain peptides achieved by a 1.7μm BEH, 100μm x 100mm C18 column either with or without pre-separation through a 5μm Symmetry, 180μm x 20mm C18 column. Chromatographic resolution was measured by calculating peak capacity, based on the time difference between the last and first eluting peptide divided by peak width at 10% peak height. Peak capacities of 95 were achieved in the presence of the pre-column, and 180 in its absence. Accurate Mass Retention Time Pairs (AMRT) were extracted from multiplexed LC-MS datasets (LC-MS^E), and only reproducible AMRTs were taken forward for data evaluation. The influence of chromatographic resolution on the number of AMRTs is shown in Figure 1 and on the number of proteins identifiable from a single sample in Figure 3. The low peak capacity detected 7,444 (Figure 1a) and the high peak one 12,636 (Figure 1b) AMRTs. All of the latter showed a relative standard deviation (RSD) of <5% of their ion intensity, and 95% showed a RSD of <0.8% of their retention time (Figure 1c). A comparison of log intensities of the two replicates from the high peak capacity separation is shown in Figure 1d as a criterion to assess run to run reproducibility.

Peak capacity was improved up to 800 in a routine experiment by substituting 1.7μm BEH 75μm x 250mm nanoUPLC colums (Waters, Milford, US) and longer gradient times. Other studies have shown that separation capacity based on 3μm particles can be improved by either lengthening the column or by modifying the column temperature and/or gradient elution profiles (Shen 2005; Wang, X. et al. 2006), however, real improvements require very long columns and/or extreme gradients, and these are rather impractical in the context of high throughput analysis.

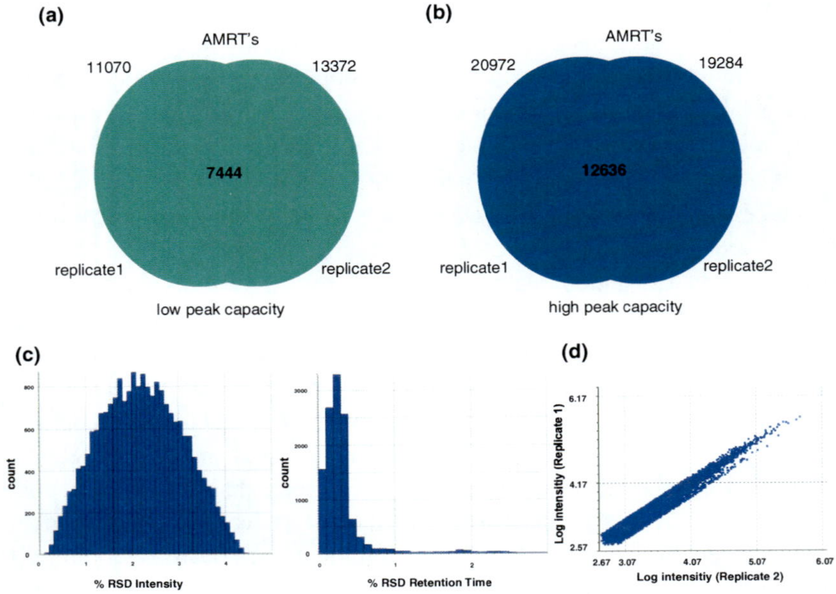

Figure 1: Accurate Mass Retention Time Pairs (AMRT) were extracted from multiplexed LC-MS datasets. AMRTs with a replication of 2 out of 2 analyses were used for further analysis. (a) 7,444 reproducible AMRTs were detected following low peak capacity separation, and (b) 12,636 reproducible AMRTs following high peak capacity separation. (c) Clustering of the latter AMRTs produced an RSD of <5% in ion intensity, and 95% produced an RSD of <0.8% in retention time. (d) A comparison of log intensities for the two replicates from the high peak capacity separation.

Two-dimensional LC-separation prior to MS analysis was introduced to the field of proteomics nearly ten years ago. At this time, a multidimensional LC system was available, based on the sequential packing of RP and strong cation-exchange particles in a biphasic column (Wolters et al. 2001; Washburn et al. 2002). The combination of multidimensional LC separation and ESI-MSMS detection was named "multidimensional protein identification technology" (Mudpit). The peak capacity of the 2-D separation is represented by the sum of the peak capacities of the individual one-dimensional methods, and can be further increased by taking advantage of more recent developments in separation technology. Recently, a comparison was made between the proteomes of Shigella dysenteriae as derived by label free LC and label free 2-D gel electophoresis (Kuntumalla et al. 2009). The former was performed by combining off-line fractionation on an ion-exchange column with RP chromatography and on-line MS detection. The MS data sets were evaluated using APEX ("absolute protein expression index") software (Vogel and Marcotte 2008), confirming the versatility of this method, which is based on computationally modified spectral counting (Vogel and Marcotte 2008; Kuntumalla et al. 2009). A comprehensive analysis using two-dimensional HPLC in combination with tandem MS was recently described for Schizosaccharomyces pombe, leading to the identification of some 3,400 proteins per sample. Spectral counting was applied for a semi-quantitative assessment of the MS data (Brill et al. 2009).

Coupling RP chromatography with a second dimension RP chromatography at a different pH (2D-RPRP LC) was proposed, with the intention of exploiting the reproducibility and separation performance of RP chromatography in both separation dimensions (Gilar, M. et al. 2005). Even if RP chromatography of peptides at contrasting pHs is only a partially orthogonal method, the improvement in separation performance and loading capacity appears to be considerable. A commercial hardware platform for 2D RP-RP nanoUPLC (2D nanoAcquity) has been recently introduced by Waters, and early outcomes based on this technology are expected in the near future.

Computational Methods for Comparative Profiling

The large increase in throughput permitted by current MS devices and the quantity of data generated by each proteomics experiment requires the development of advanced computational and statistical methods for evaluation. Here we focus on computational procedures appropriate for the analysis of label-free LC-MS data sets, where the major priority is the detection, matching and alignment of chromatographic peaks. Which particular statistical evaluation of experimental data is required depends on the design of the biological experiment and the instrumental setup that was used for the analysis. Simple quantitative proteomics experiments may only require the estimation of mean values with their associated standard errors, while clustering approaches to deal with the type of dataset generated from a kinetic study requires more sophisticated computational methods. Specific database search algorithms are vital for protein identification, as discussed in greater detail in the following section (“Databases and Search Algorithms”). A number of computational procedures aimed at the comparative quantification of proteins derived from label-free LC-MS experiments have been published and discussed in recent years (Wong et al. 2007; America and Cordewener 2008; Kumar and Mann 2009; Roewer et al. 2009) ; some of these are open-source and others are only available on a commercial basis (Table 2). Some are still under development, but a number are fully operational harbouring more or less clear user interface components (e.g. MetAlign, MSight, MsInspect, PePPeR, SuperHirn, VIPER), that are generally better evolved in commercially available software tools (e.g. SIEVE and PLGS). The commercially available software packages have been deliberately designed to process raw data produced by a specific MS device; thus, their performance is superior to the more generic packages, but they cannot be readily be used with non-recommended MS instrumentation. As a result, there has been a push to assemble a suite of freeware, in particular the Trans-Proteomic Pipeline (TPP) developed by the Seattle Proteomics Centre (SPC) (http://tools.proteomecenter.org/software.php). Most freeware packages use the mzXML format for the import of raw data, so data transformation is generally needed – a process which is time consuming and also increases data volume considerably (by 3-5 fold). The Proteomics Standards Initiative has been attempting to develop mzML, a file format which is expected to gradually replace existing ones.

The first steps in data processing are peak detection and peak alignment. Various peak detection algorithms have been described, and a comprehensive discussion of the alternatives has been published earlier (Listgarten and Emili 2005). The key steps are noise filtering, background subtraction, peak detection and grouping over multiple consecutive MS acquisitions for each individual peak. The best peak detection results are thought to be generated by algorithms which take both m/z and the time dimension into account (Du et al. 2007). However, a strong prerequisite for robust peak detection is the quality and reproducibility of the LC-MS acquisition (i.e., the high resolution and retention time stability assured by LC separation, together with a high accuracy of mass estimation). The next data processing step is the alignment of peaks across the dataset as a whole. Many algorithms are in use for matching peaks on the basis of mass (or m/z and charge state), retention time and/or intensity (America and Cordewener 2008), but fully rely on LC-MS-only datasets. A more recently developed method, which combines LC-MS data with fragment information, relies either on rapid switching between the precursor and fragment ion mode (modern ion trap and Q-TOF instruments) or on (semi)parallel acquisition (ion trap FT-MS or ion trap-Orbitrap instruments). The algorithms switch either between MS and MS/MS mode (data dependent acquisition) or between low and high collision energy MS (multiplexed MS, data independent acquisition). During MS/MS acquisition, precursor ions are filtered and information relating to co-eluting precursor ions is lost. The selection of precursor ions is a biased process, which is necessarily limited by the overall number of selectable precursors present in the sample. A multiplexed LC-MS experiment is unbiased, and provides information about all precursor and fragment ions present. Once the problem of low initial specificity in a multiplexed LC-MS experiment has been overcome, this mode of data acquisition is superior with respect to both the qualitative and quantitative analysis of complex mixtures (Geromanos et al. 2009). All processing algorithms need to align LC-MS(/MS) data (or the multiplexed LC-MS data) according to a time-frame determined by the behaviour of well characterized peptide sequences (Fischer et al. 2006; Jaffe et al. 2006; Prince and Marcotte 2006; Silva et al. 2006; Vissers, J.P. et al. 2007). PLGS is the only software package to date able to handle multiplexed LC-MS data.

Further data processing - in particular data normalization and filtering - is needed for the extraction of meaningful information. Normalization can be based on mass (m/z or MW values), retention time and/or peptide abundance (peak intensity or area), and is essential for a comparative analysis of the output of multiple LC-MS separations. A comprehensive summary of the relevant issues has been given by (America and Cordewener 2008). Filtering options are intended to facilitate the detection of low quality spectra in terms of minimal peak intensity, quality of isotope pattern, consistency of charge detection and chromatographic elution shape prior to peak alignment. The output of these various alignment procedures is in the form of a table listing the matched items; in PLGS and VIPER, this is referred to as the "accurate mass retention time" (AMRT) table, while in MsInspect and SpecArray, it is called "peptide array". Some software packages do not progress beyond this point, leaving all subsequent data analysis to be performed by other software. However, most packages do provide options for subsequent statistical analysis, comparative analysis and data visualization. Various data evaluation parameters can be taken into account, including isotope distribution, retention time drift during alignment, and intensity variation between replicates. Visual inspection tools can also help to assess the quality of the processed data, while even basic statistical evaluation (such as the calculation of standard errors) is obligatory if an objective view of the quality of data processing and experimental accuracy is required.

Provided that the match table is of sufficient quality, a comparative analysis between individual samples or groups of samples can be performed. The strategies available for label-free protein quantification experiments can be divided into those based on peptide identification before quantification, and those which rely on precursor ion information alone. In theory, the ion abundance is reflected by the height or area of a peak at a specific m/z, (i.e., the number of ions detected by the MS device at any given time). Therefore, the result can be visualized as a two-dimensional image with retention time (x-axis), *m/z* (y-axis), together with the ion intensity. Consequently, image based software, such as MSight, 2DICAL, and MatchRx (which rely directly on image analysis) or XCMS, SpecArray, msInspect, and OpenMS (which detect peptide features and extract quantitative information directly from the raw data), are appropriate for comparative analysis (see Table 2). A drawback of this approach is the dependency of the ionization process on the molecular composition of the target molecules. Thus, data derived from two different precursor ions cannot be directly compared with one another without taking into account differences in ion composition, as explored by a number of investigators (Fischer et al. 2006; Prince and Marcotte 2006; Wong et al. 2007). However, by initially applying database search tools (Table 3) to identify relevant peptides, quantitative information relating to those peptides can be extracted from the raw data. Some recent computational methods incorporating this approach include MSQuant, Serac PeakExtractor, ASAPratio, XPRESS, PLGS and Census (Table 2). These allow for the extraction of peptide intensities, following their identification. Combining ion intensity data from peptides derived from a single protein gives a means to estimate error, and thus increases the statistical confidence of intensity comparisons in multiple LC-MS data sets. Other software tools have been developed to address spectral counting. The spectral count of a given protein refers to the number of MS/MS spectra acquired from its proteolytic peptide ions during an individual LC-MS/MS run; frequently the more abundant peptide(s) will be selected for MS/MS analysis (Liu et al. 2004).

Table 2. Available software packages relevant for protein/peptide quantification from label-free LC-MS experiments adapted from (America and Cordewener 2008)

Software	Website	Availability
Quantification software with full functionality[a]		
PLGS Identity[E] Expression Informatics[d, l]	http://www.waters.com/	commercial
SIEVE[d, l, v]	http://www.thermo.com/	commercial
DeCyderMS[l, v]	http://www.gelifesciences.com/	commercial
Rosetta Elucidator[l, v]	http://www.rosettabio.com/products/elucidator/default.htm	commercial
MassView[l]	custom	upon request
Viper[d, l, v, rd]	http://ncrr.pnl.gov/software	open source
OpenMS[d, l, v, rd]	http://www.openMS.de	open source
MS-Xelerator	http://www.msmetrix.com	commercial
SuperHirn[d, v]	http://tool.proteomecentre.org/SuperHirn.php	open source
CRAWDAD[l, rd, v]	http://proteome.gs.washington.edu/software/crawdad	upon request
2DICAL[l, v,]	custom	upon request
Other quantification software		open source
MzMine[v]	http://mzmine.souceforge.net/index.shtml	open source
msInspect[v, l, rd]	http://proteomics.fhcrc.org/CPL/msinspect.html	open source
SpecArray[v]	http://tools.proteomecentre.org/SpecArray.php	open source
PEPPeR[l]	http://www.broad.mit.edu/cancer/software/genepattern/desc/proteomics.html	open source
ProtQuant	http://www.agbase.msstate.edu/	free for academic use
MatchRx[v]	custom	no information
ASAPratio	http://tools.proteomecenter.org/software.php	open source
XPRESS	http://tools.proteomecenter.org/software.php	open source
Alignment software		
MetAlign	http://www.metalign.wur.nl	free for academic use
MSight[v]	http://www.expasy.org/MSight/	open source
Xalign[d]	Zhang100@purdue.edu	upon request
CHAMPS	http://www.pasteur.fr/recherche/unites/Biolsys/champs/index/htm	web server
OBI-WARP[l]	http://bioinformatics.icmb.utexas.edu/obi-warp/	open source
LCMSWARP[d]	http://ncrr.pnl.gov/software	open source
LCMS2D[v]	http://www.bioc.aecom.yu.edu/labs/angellab/	freeware
PETAL[d]	http://peiwang.fhcrc.org/research-project.html	open source

Table 2. (Continued)

Software	Website	Availability
CPM[l, d]	http://www.cs.toronto.edu/~jenn/CPM/	free for academic use
MSQuant[v]	http://msquant.sourceforge.net	open source
Census[l, v]	http://fields.scripps.edu/census	freeware
XCMS[v]	http://masspec.scripps.edu/xcms/xcms.php	open source
Spectral counting related software		
PeptideProphet	http://tools.proteomecentre.org/PeptideProphet.php	open source
ProteinProphet	http://tools.proteomecentre.org/ ProteinProphet.php	open source
NoDupe	http://fields.scripps.edu/nodupe	freeware

[a] Functionality includes: peak/feature detection, de-isotoping, batch processing, alignment, result visualization, and statistical analysis

[l] Link MS to MS/MS or to corresponding fragment data information

[v] Software contains 1-D/ or 2-D visualization of LC-MS data

[d] Database environment

[rd] Results database

Software tools implementing spectral counting (e.g., PeptideProphet and ProteinProphet) are designed to initially interpret all MS/MS spectra via a database search algorithm, and then to calculate the peptide/protein counts based on all spectra associated with a particular peptide ion or protein. An alternative approach, such as that used by NoDupe, clusters mass spectra based on their similarity prior to a database search and quantitative analysis, aiming to minimize the need to scan similar spectra.

All the methods described above have been successfully applied to proteomics experiments (Wong et al. 2007; America and Cordewener 2008; Roewer et al. 2009). Integrated software tools supporting the analysis of various analytical approaches, however, remain rare. The quantitative Census tool was released recently, which is compatible with data-independent and single-reaction monitoring analyses derived from both label-free and labelling experiments, and is able to perform quantitative analyses based on spectral counting. (Park et al. 2008). It relies on an LC-MS peak area approach to align chromatograms, and supports several input file formats. The program is available free of charge for academic and non-profit use (Table 2). Software packages combining data processing, data evaluation, peptide/protein identification and quantification remain limited to commercial products.

Multivariate Statistics for Comparative Proteomics and Kinetic Analyses

In principle, quantitative output from typical MS datasets can be analysed using multivariate statistics, such as those currently applied to microarray experiments. However, to date the few examples of the application of label-free quantitative kinetic profiling on the protein level present in the literature show that the procedure cannot be considered as routine. Kinetic profiles of specific proteins are sometimes visualized by intensity plots derived from expression under a range of conditions (Cheng et al. 2009). Multivariate statistical analysis of LC-MS data is typically restricted to a principal component analysis (PCA) and/or hierarchical clustering. PCA is appropriate where the establishment of relationships between sample groups (classes) is the aim (Gaspari et al. 2006; Kempermann et al. 2007; Roewer et al. 2009). In a comprehensive study label-free proteomics was applied to the quantitative analysis of five mouse core proteomes (Cutillas and Vanhaesebroeck 2007), where the visualization and evaluation of LC-MS data was performed by means of a hierarchical cluster analysis using Cluster and Treeview (Eisen et al. 1998). The 1,000 most abundant proteins were grouped by complete linkage clustering, based on the Pearson correlation similarity metric. Variations of PCA, such as hierarchical PCA or independent component analysis (ICA) have also been applied to interpret quantitative experimental data derived from label-free LC-MS measurements. In a study investigating the response of A. thaliana to abiotic temperature stress, an ICA-based comparison between a wildtype and a mutant proteome was able to resolve a number of genotype-dependent responses to the temperature treatment, as well as certain genotype-independent temperature compensation mechanisms (Wienkoop et al. 2008).

Most regulatory mechanisms are non-linear, which is a problem for standard PCA (Seiffert et al. 2005). To analyze correlations across several data sets while aiming at unravelling co-regulation and/or kinetic patterns, computational intelligence based clustering algorithms, such as Self-Organizing Maps (SOM) and Neural Gas (NG) are most suitable. Similarities between samples provide the basis for clustering, context-specific data representation, dimensionality reduction by data projection and embedding, etc. When utilizing these advanced techniques, the conducted data analysis benefits from the ability to utilize non-standard similarity measures that often reflect the underlying biological system, e.g. biological networks, much better than commonly used standard metrics, such as Euclidean distance, Manhattan distance, etc. This way it becomes possible to apply application-driven similarity measures that even do not necessarily need to be defined analytically (in terms of mathematical equations). Moreover, many of these techniques keep the data within their natural physical domain. As a result, the interpretation of clusters and the relationship between them becomes easier and more intuitive. SOM clustering combined with Pearson correlation distance scoring was applied for 2-D gel image analysis in order to identify clusters of proteins which are co-regulated in *Arabidopsis thaliana* under various light regimes (Kim et al. 2006). Other studies using computational intelligence based clustering algorithms for data evaluation are rare or simply missing.

In our own experiments, we have sought to define the proteome of the barley caryopsis and to identify proteins which are co-regulated during its development. For this purpose barley seeds of various developmental stages (3, 5, 7, 10, and 16 days after flowering) were analysed. The workflow involved in a label-free comparative multiplexed LC-MS experiment is presented in figure 2. Whole crude extracts were digested and tryptic peptides analysed directly using a nanoLC system combined with ESI-Q-TOF MS (Waters). Data acquisition was performed by a data independent multiplexed LC-MS strategy using fast alternate switching between low and high energy mode. PLGS IdentityE Expression Informatics software (Waters) was used for data processing and protein identification, processing the intensities of molecular ions for quantification, and identifying the fragments and molecular ions. The PLGS2.4 Expression module can generate and cluster AMRTs from multiplexed LC-MS data sets without any prior protein identification. A comparison of log intensities for two replicates from a typical no pre-column separation is shown as Figure 1d. Every LC-MS peak can be assigned an accurate mass and retention time. The number of AMRTs identified per run depends heavily on the noise threshold level. Thus, clustering and comparison of the AMRTs from two or more independent measurements provides an efficient means of reducing background noise. Three replicate injections under nearly identical conditions appeared to be sufficient to obtain reliable quantification. Drifts in retention time could be minimized by recycling the same column, and applying the identical gradient and temperature conditions. As many as 70,000 significant AMRT clusters were detected, too many to full identify given the limits of genomic information available. Quantification between any two samples can be performed at either the peptide or protein level (Silva et al. 2005), in which quantification at the protein level involves mapping of detected peptides to proteins in the database. Besides, quantification at the peptide level allows also groups of unidentified peptides. For an elucidation of biologically related statistically significant and objective kinetic patterns and biomarker identification, multivariate statistics was applied to the detected AMRTs, which can be identified afterwards. The data were exported, pre-processed (missing value elimination, averaging, and normalization) and an initial visualization performed to ensure data quality and the appropriateness of the clustering algorithm. The AMRTs were then clustered using Neural Gas (Martinetz and Schulten 1991), on the basis of how expression was affected by development. Each peptide was attributed to a cluster representing the closest related prototype (among ten selected a priori) according to its intensity profile. Peptide identification relied on the database search algorithm implemented in PLGS IdentityE Expression Informatics software. The resulting clusters were evaluated manually with respect to their biological meaning.

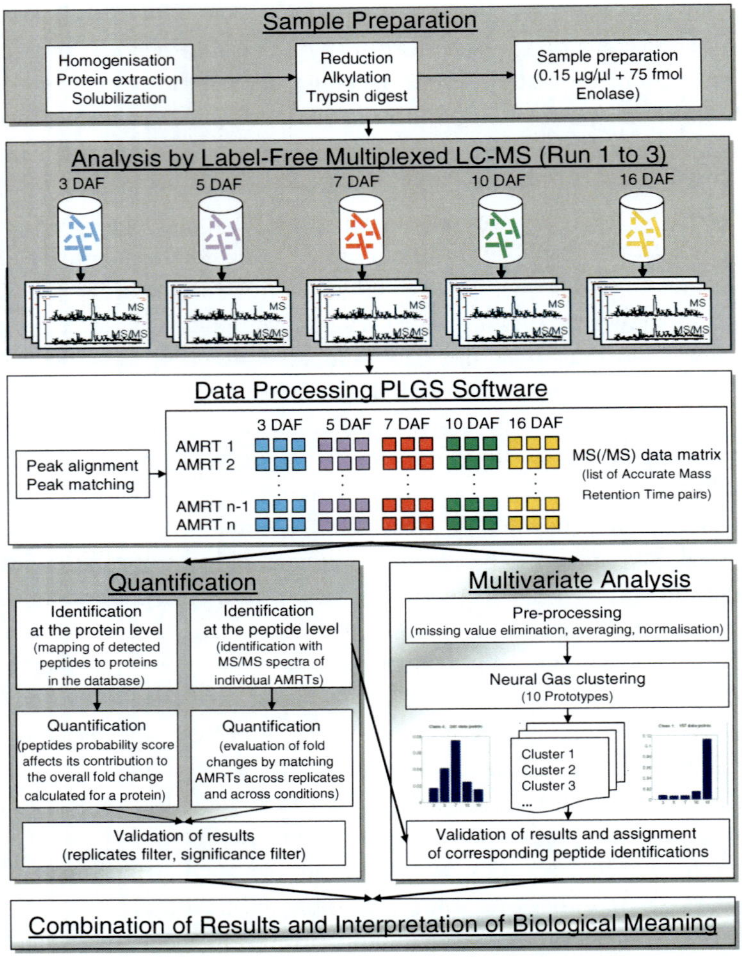

Figure 2. A typical workflow involved in a label-free comparative multiplexed LC-MS experiment. Barley caryopses at 3, 5, 7, 10 and 16 days after flowering were homogenized, and soluble proteins extracted using TCA/acetone. The protein pellet was treated with RapiGest SF (Waters) and proteins were reduced, acylated and digested. Peptides were separated on a nanoLC-system and data acquisition achieved using a data independent strategy, with fast alternate switching between low and high energy mode. Three independent LC runs were performed per harvest time. Results from multiplexed LC-MS experiments were initially scanned using PLGS2.4 software based on AMRT signatures, and the resulting AMRTs subsequently clustered using the NG method. Protein and peptide identifications were performed using database search algorithms implemented in PLGS software. Finally, the results were combined and biologically interpreted.

DATABASES AND SEARCH ALGORITHMS

The majority of tools described above have been designed to extract peptide intensities without assuming any prior knowledge of peptide identity. Moreover, in many cases quantitative data are determined for peptides that cannot be identified so far. To make use of the information obtained by the progress in protocols and instrumentation, a completely sequenced genome is essential for many of the proteomics experiments performed nowadays. While a completely sequenced genome implies that every gene product can be identified in principle, only a few eukaryotic organisms have as yet been completely sequenced (*http://img.jgi.doe.gov/cgi-bin/pub/main.cgi?section =FindGenomes&page=findGenomes*). The highest number of completely sequenced genomes is available for smaller genomes from Viruses, Archebacteria and Bacteria. The complete sequencing of animal genomes has been so far focusing on important model organisms for research. Only few animal genomes have been completely sequenced or are available as draft assemblies, among them human, mouse, rat, dog, cattle, guinea pig, zebrafish, xenopus, honeybee, drosophila, and mosquito. In the case of plants complete genome reference sequence information is only available for the two model plants *Arabidopsis thaliana* and rice, although the annotation of the Arabidopsis genome is currently at a more advanced stage than that of rice (Jorrin-Novo et al. 2009). Thus, proteomics experiments with other plant species remain difficult, although this situation is likely to improve in the near future. For poplar the complete genome sequence is now available (*Populus trichocarpa*) and the genome sequences of several other species, such as barrel medic and maize (Haynes and Roberts 2007), as well as of barley (Schulte et al. 2009) are scheduled to be completed in the next few years. In the meantime, extensive EST databases and combined information from various databases can provide an alternative to the full genome sequence (Witzel et al. 2007; Brumbarova et al. 2008; Hoehenwarter et al. 2008; Lippmann et al. 2009). While cross-species orthology has been shown to be effective in mammalian proteomic analyses (Vissers, J.P.C. et al. 2009), it is not very likely to be the same in plants as they have to express a high number of specialized proteins for their extended secondary metabolism.

A number of database search tools are available for protein identification, and quantitative data for known peptides can be extracted from the raw data (see Table 3). The reliability of identification depends heavily on the sensitivity and mass accuracy of the MS analysis. Thus, bioinformatics tools are needed to deal with the problem of false positives. Attempts to develop these have been ongoing for a number of years; the commonest approach has been to combine a normal 'forward' database search with a sequence-reversed or sequence-randomized 'target-decoy' database (Elias and Gygi 2007).

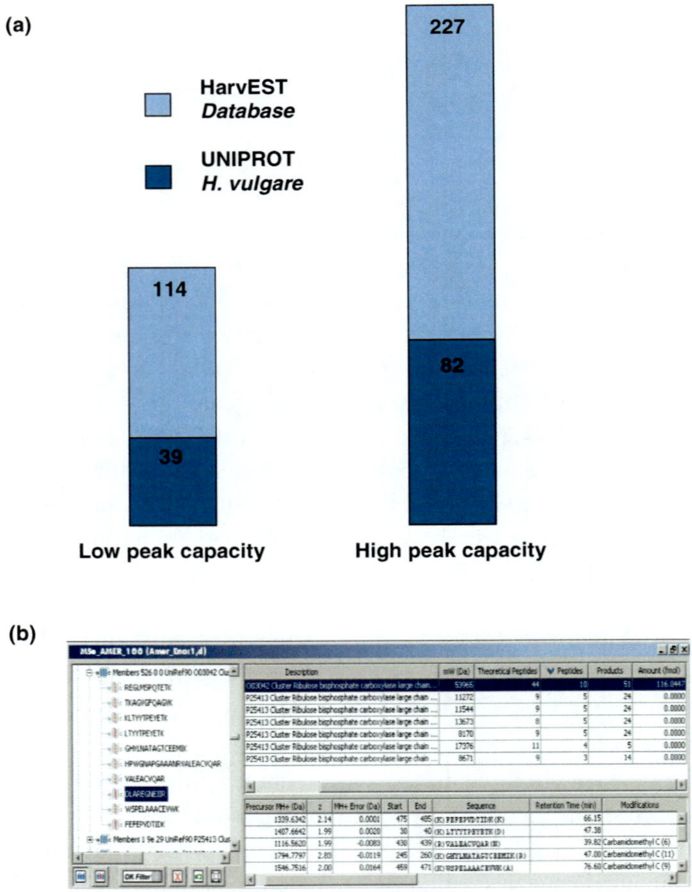

Figure 3. (a) Protein identification was effected using PLGS2.4, and the results were stringently filtered (replication 2 out of 2 analyses, no homologue protein identifications, false discovery rate on protein level <1%). A larger number of proteins were identified from the HarvEST database than from UNIPROT, and when applying higher chromatographic resolution. (b) Raw identification list for RuBisCo large chain shows a subset of homologous peptides shared by various HarvEST entries. As all peptides from any of these entries could be explained to be a peptide of the first entry (O03042), PLGS2.4 assigned the whole protein amount to this first entry.

As a result, search algorithms implementing the estimation of false discovery rates have been developed, such as the PeptideProphet and ProteinProphet software (Keller et al. 2002; Nesvizhskii et al. 2003), and the ProteinLynx GLOBAL Server software (PLGS, Waters, Manchester, UK).

The only currently commercially available system which combines raw data processing of multiplexed LC-MS data sets with protein/peptide identification and label-free quantification is PLGS. The quality of the database being searched is clearly also a critical issue. Here, the most significant problems are the level of redundancy and the lack of a search algorithm capable of distinguishing between protein isoforms. This latter problem has been addressed in PLGS2.4, which relies on the extensive sequence coverage gained in multiplexed LC-MS experiments, the intrinsic quantitative information about every identified peptide, and a reliable annotation of known homologues and the unique peptides which define protein isoforms.

PLGS2.4 and various plant databases were used to identify barley developing caryopsis proteins. Given that the barely genome sequence is not yet complete, searches had to be based on all 2577 barley protein entries in UNIPROT (July 2009) and a set of translated ESTs (archived in HarvEST), where all predicted proteins with >90% sequence homology were collapsed into a single entry. Depending on the quality of the chromatographic resolution, a set of 39-82 proteins (2/2 replications, FDR on protein level <1%, no homologues) were recovered from the UNIPROT database, underlining the incompleteness of this database. The HarvEST ESTs, which still include nine variants of the RuBisCo large chain, produced 114-227 proteins without redundant annotations (Figure 3a), which still included nine variants of the RuBisCo large chain. Since PLGS2.4 can distinguish between homologues and protein isoforms, redundancies could be removed effectively (Figure 3b).

Table 3. Available software relevant for protein identification from label-free LC-MS/MS experiments [adapted from (Wong et al. 2007)]

Software	Website	Availability
GutenTag	http://fields.scripps.edu/Guten Tag	Freeware
Inspect	http://peptide.ucsd.edu/inspect.html	Open source
MASCOT	http://www.matrixscience.com	Commercial
PLGS	http://www.waters.com	Commercial
Sequest	http://fields.scripps.edu/sequest	Commercial
X!Tandem	http://www.thegpm.org	Open source

PERSPECTIVES

Current proteomics techniques allow the analysis of thousands of proteins in a wide variety of organisms and biological samples. The widespread use of proteomics has emphasized the need for both standardized forms of reporting and improved bioinformatics tools (Moxon et al. 2009). Continuing improvements in proteomics technology, particularly with respect to LC-based separation (e.g., UPLC, 2-D LC) and MS detection has served to increase both the number of protocols and data sets. A common format for the description of experiments and for reports of the data output is now needed. A first attempt to achieve this was the concept of the "Minimum Information About a Proteomics Experiment" (MIAPE), suggested by the HUPO Proteomics Initiative (Taylor et al. 2007). Reporting the detail of workflows from sample preparation through experimental setup to data analysis can only strengthen the value of applying proteomics to address biological questions. The protocols developed by the plant metabolomics community (de Vos et al. 2007) represent a viable role model. Some protocols for quantitative protein profiling by MS using label-free techniques have recently been made available (Haqqani et al. 2008; Wisniewski et al. 2009).

Integrated software solutions to lighten data handling are a second priority. First efforts combining all the important steps into single software packages have been taken. Some of these will necessarily be platform-specific, such as the PLGS IdentityE Expression System, as this is part of an integrated system solution (hardware and software). Others will need to be designed to handle data acquisition formats from different vendors. The commercially available Progenesis LC-MS software (Nonlinear Dynamics, Newcastle, UK) allows the visualization and statistical analysis of differential expression. The main problem in developing complete software solutions is how to combine raw data derived from different platforms (Mortensen et al. 2009). However, the development of the mzML format should help in the future, as most of the technology suppliers have now agreed to use it (Deutsch 2008).

A third area for improvement lays in the estimation accuracy of peptide intensity values, together with separation reproducibility. Recent successes in increasing the performance of MS devices should aid the comparative analysis of complex peptide mixtures, but they also underline the importance of improving retention time stability.

The protein identification process remains a weak link in the technology. Although the growing analytical capacity of MS systems is generating ever larger data sets, the biological significance of these proteins rests on being able to identify them. Protein databases remain by and large incomplete, especially for species where even the complete genome sequence is as yet unfinished. Redundancy remains a particular problem, although there are a number of current efforts to combine databases. One such is the Universal Protein Resource (Uniprot), which collects and curates entries for several species and has initiated a specific Plant Proteome Annotation Program (PPAP) to provide a centralized and authoritative source of information (Schneider et al. 2005). A second current initiative, the Plant Proteomics Database (PPDB), which initially was restricted to plastid entries, is now being regularly updated and curated (Sun et al. 2008). ProMEX (*http://promex.mpimp-golm.mpg.de/cgi-bin/peplib.pl*) is a mass spectral reference library for plants (Hummel et al. 2007). The IPI (international protein index, *http://www.ebi.ac.uk /IPI/IPIhelp.html*) database combines information from a number of model organisms (human, mouse, rat, zebrafish, arabidopsis, chicken, cow), aiming to "effectively maintain a database of cross references between the primary data sources while providing minimally redundant yet maximally complete sets of proteins". Inter-relationships between protein families, based on both structure and/or function represents a higher level of complexity (Wu et al. 2004). Several in silico methods aimed at the identification of protein function are already in the public domain (e.g., iProClass: *http://pir. georgetown.edu/iproclass/*, GeneGo: *http://www.genego.com/*, iSpider: *http://www. ispider.manchester.ac.uk /cgi-bin/ispider.pl*).

Beyond this lies the connection between the metabolome and the proteome (Hennig 2007; Weckwerth 2008; Wienkoop et al. 2008; Bylesjö et al. 2009; Lippmann et al. 2009). In the future, integrated analyses along these lines will underpin system-wide biology. This scenario demands the development of ever more sophisticated software tools, which greatly rely on the will of data sharing and unique output formats, for data handling, protein identification and protein integration together with the setup of integrative databases such as developed for genomic approaches (e.g. Genevestigator). In parallel development and extension of tools for visualisation of biological networks, both on the level of metabolites and of enzymes needs to be continued such as MapMan (Usadel et al. 2005; Usadel et al. 2009).

CONCLUSIONS

Much progress has been made in analytical instrumentation, along with the methods needed for quantitative proteomic data analysis. Enhancements in the stability and accuracy of both peptide separation and mass detection have enabled the development of label-free quantitative analysis. Experimental reproducibility is paramount, yet it has not been widely accepted as such as yet. As experimental variation can be caused by biological sample, by the instrumentation as well as by the operator, quantitative proteomics experiments need to include both biological and technical replicates. Statistically meaningful comparisons require reproducible peptide and protein identification. Generating meaningful information from experimental data requires appropriate bioinformatics and statistical analyses. Although a start has been made to address many of these issues, there is a pressing need to develop fully functional, robust, user-friendly software. The entire workflow, from sample isolation to statistical evaluation needs to be borne in mind when choices are being made as to which proteomics platform is appropriate.

REFERENCES

Addona, T.A., Abbatiello, S.E., et al. (2009). "Multi-site assessment of the precision and reproducibility of multiple reaction monitoring-based measurements of proteins in plasma." *Nat Biotech* 27(7): 633-641.

America, A.H.P. and Cordewener, J.H.G. (2008). "Comparative LC-MS: A landscape of peaks and valleys." *Proteomics* 8: 731-749.

Bachi, A. and Bonaldi, T. (2008). "Quantitative proteomics as a new piece of the systems biology puzzle." *Journal of Proteomics* 71(3): 357-367.

Bodenmiller, B., Mueller, L.N., et al. (2007). "Reproducible Isolation of Distinct, Overlapping Segments of the Phospho-Proteome." *Nature Methods* 4: 231-237.

Brill, L.M., Motamedchaboki, K., et al. (2009). "Comprehensive proteomic analysis of Schizosaccharomyces pombe by two-dimensional HPLC-tandem mass spectrometry." *Methods* 48: 311-319.

Brumbarova, T., Matros, A., et al. (2008). "A proteomic study showing differential regulation of stress, redox regulation and peroxidase proteins by iron supply and the transcription factor FER." *The Plant Journal* 54(2): 321-334.

Bylesjö, M., Nilsson, R., et al. (2009). "Integrated Analysis of Transcript, Protein and Metabolite Data To Study Lignin Biosynthesis in Hybrid Aspen." *Journal of Proteome Research* 8(1): 199-210.

Cheng, F.-Y., Blackburn, K., et al. (2009). "Absolute Protein Quantification by LC/MSE for Global Analysis of Salicylic Acid-Induced Plant Protein Secretion Responses." *Journal of Proteome Research* 8(1): 82-93.

Cooper, J.W., Wang, Y., et al. (2004). "Recent advances in capillary separations for proteomics." *Electrophoresis* 25: 3913-3926.

Cutillas, P.R. and Vanhaesebroeck, B. (2007). "Quantitative Profile of Five Murine Core Proteomes Using Label-free Functional Proteomics." *Molecular & Cellular Proteomics* 6: 1560-1573.

de Godoy, L.M.F., Olsen, J.V., et al. (2008). "Comprehensive mass-spectrometry-based proteome quantification of haploid versus diploid yeast." *Nature* 455: 1251-1254.

de Vos, R., Moco, S., et al. (2007). "Untargeted large-scale plant metabolomics using liquid chromatography coupled to mass spectrometry." *Nature Protocols* 2(4): 778-791.

Deutsch, E. (2008). "mzML: A single, unifying data format for mass spectrometer output." *Proteomics* 8: 2776-2777.

Du, P., Sudha, R., et al. (2007). "Data reduction of isotope-resolved LC-MS spectra." *Bioinformatics* 23: 1394-1400.

Eisen, M.B., Spellman, P.T., et al. (1998). "Cluster analysis and display of genome-wide expression patterns." *PNAS* 95: 14863-14868.

Elias, J.E. and Gygi, S.P. (2007). "Target-decoy search strategy for increased confidence in large-scale protein identifications by mass spectrometry." *Nature Methods* 4: 207-217.

Fang, R., Elias, D.A., et al. (2006). "Differential Label-free Quantitative Proteomic Analysis of Shewanella oneidensis Cultured under Aerobic and Suboxic Conditions by Accurate Mass and Time Tag Approach." *Molecular & Cellular Proteomics* 5: 714-725.

Fischer, B., Grossmann, J., et al. (2006). "Semi-supervised LC/MS alignmentfor differential proteomics." *Bioinformatics* 22: 132-140.

Gaspari, M., Verhoeckx, K.C.M., et al. (2006). "Integration of Two-Dimensional LC-MS with Multivariate Statistics for Comparative Analysis of Proteomics Samples." *Anal. Chem* 78(7): 2286-2296.

Geromanos, S.J., Vissers, J.P.C., et al. (2009). "The detection, correlation, and comparison of peptide precursor and product ions from data independent LC-MS with data dependant LC-MS/MS." *Proteomics* 9: 1683-1695.

Gilar, M., Olivova, P., et al. (2009). "Comparison of 1-D and 2-D LC MS/MS methods for proteomic analysis of human serum." *Electrophoresis* 30(7): 1157-1167.

Gilar, M., Olivova, P., et al. (2005). "Two-dimensional separation of peptides using RPRP-HPLC system with different pH in first and second separation dimensions." *Journal of Separation Science* 28: 1694-1703.

Gingras, A.-C., Aebersold, R., et al. (2005). "Advances in Protein Complex Analysis using Mass Spectrometry." *Journal of Physiology* 563(1): 11-21.

Gobom, J., Nordhoff, E., et al. (1999). "Sample Purification and Preparation Technique Based on Nano-scale Reversed-phase Columns for the Sensitive Analysis of Complex Peptide Mixtures by Matrix-assisted Laser Desorption/Ionization Mass Spectrometry." *Journal of Mass Spectrometry* 34(2): 105-116.

Haqqani, A.S., Kelly, J.F., et al. (2008). Quantitative protein profiling by mass spectrometry using label-free proteomics. *Methods in Molecular Biology*. M. Starkey and R. Elaswarapu. Totowa, NJ, Humana Press. Genomics Protocols: 241-256.

Haynes, P.A. and Roberts, T.H. (2007). "Subcellular shotgun proteomics in plants: Looking beyond the usual suspects." *Proteomics* 7: 2963-2975.

Hennig, L. (2007). "Patterns of beauty – omics meets plant development." *TRENDS in Plant Science* 12(7).

Hewel, J.A. and Emili, A. (2008). "High-resolution biomarker discovery: Moving from large-scale proteome profiling to quantitative validation of lead candidates." *Proteomics Clin. Appl.* 2: 1422-1434.

Hoehenwarter, W., van Dongen, J.T., et al. (2008). "A rapid approach for phenotype-screening and database independent detection of cSNP/protein polymorphism using mass accuracy precursor alignment." *Proteomics* 8: 4214-4225.

Huges, M.A., Silva, J.C., et al. (2006). "Quantitative proteomic analysis of drug-induced changes in mycobacteria." *Journal of Proteome Research* 5: 54-63.

Hummel, J., Niemann, M., et al. (2007). "ProMEX: a mass spectral reference database for proteins and protein phosphorylation sites." *BMC Bioinformatics* 8: 216.

Jaffe, J.D., Mani, D.R., et al. (2006). "PEPPeR, a platform for experimental proteomic pattern recognition." *Molecular & Cellular Proteomics* 5: 1927-1941.

Jahn, O., Tenzer, S., et al. (2009). "Myelin Proteomics: Molecular Anatomy of an Insulating Sheath." *Molecular Neurobiology*.

Jensen, O. (2006). "Interpreting the protein language using proteomics." *Nature Reviews Molecular Cell Biology* 7: 391-403.

Johnson, S.A. and Hunter, T. (2004). "Phosphoproteomics finds its timing." *Nature Biotechnology* 22(9): 1093-1094.

Jorrin-Novo, J.V., Maldonado, A.M., et al. (2009). "Plant proteomics update (2007-2008): Second-generation proteomic techniques, an appropriate experimental design, and data analysis to fulfill MIAPE standards, incrase plant proteome coverage and expand biological knoledge." *Journal of Proteomics* 72(3): 285-314.

Keller, A., Nesvizhskii, A.I., et al. (2002). "Empirical Statistical Model To Estimate the Accuracy of Peptide Identifications Made by MS/MS and Database Search." *Anal. Chem.* 74(20): 5383-5392.

Kempermann, R.F.J., Horvatovich, P.L., et al. (2007). "Comparative Urine Analysis by Liquid Chromatography-Mass Spectrometry and Multivariate Statistics: Method Development, Evaluation, and Application to Ptroteinurea." *Journal of Proteome Research* 6(1): 194-206.

Kim, D.S., Cho, D.S., et al. (2006). "Proteomic pattern-based analyses of light responses in Arabidopsis thaliana wild-type and photoreceptor mutants." *Proteomics* 6: 3040-3049.

Kirkland, J.J. and De Stefano, J.J. (2006). "The art and science of forming packed analytical high-performance liquid chromatography columns." *J. Chromatogr. A* 50: 1126.

Krämer-Albers, E.-M., Bretz, N., et al. (2007). "Oligodendrocytes secrete exosomes containing major myelin and stress-protective proteins: Trophic support for axons?" *Proteomics Clin. Appl.* 1: 1446-1461.

Kumar, C. and Mann, M. (2009). " Bioinformatics analysis of mass spectrometry-based proteomics data sets." *FEBS Lett.* 583(11): 1653-1808.

Kuntumalla, S., Braisted, J., et al. (2009). "Comparison of two label-free global quantitation methods, APEX and 2D gel electrophoresis, applied to the Shigella dysenteriae proteome." *Proteome Science* 7(1): 22.

Levin, Y., Schwarz, E., et al. (2009). "Label-free LC-MS/MS quantitative proteomics for large-scale biomarker discovery in complex samples." *Journal of Separation Science* 30(14): 2198-2203.

Li, G.-Z., Vissers, J.P.C., et al. (2009). "Data searching and accounting of multiplexed precursor and product ion spectra from the data independent analysis of simple and complex peptide mixtures." *Proteomics* 9(6): 1696-1719.

Lilley, K.S. and Dupree, P. (2006). "Methods of quantitative proteomics and their application to plant organelle characterization." *Journal of Experimental Botany* 57(7): 1493-1499.

Lippmann, R., Kaspar, S., et al. (2009). "Protein and Metabolite Analysis Reveals Permanent Induction of Stress Defense and Cell Regeneration Processes in a Tobacco Cell Suspension Culture." *Int. J. Mol. Sci.* 10: 3012-3032.

Listgarten, J. and Emili, A. (2005). "Statistical and Computational Methods for Comparative Proteomic Profiling Using Liquid Chromatography-Tandem Mass Spectrometry " *Molecular & Cellular Proteomics* 4: 419-434.

Liu, H., Sadygov, R.G., et al. (2004). "A model for random sampling and estimation of relative protein abundance in shotgun proteomics." *Analytical Chemistry* 76: 4193-4201.

Mann, M. (2009). "Comparative analysis to guide quality improvements in proteomics." *Nature Methods* 6(10): 717-719.

Manza, L.L., Stamer, S.L., et al. (2005). "Sample preparation and digestion for proteomic analyses using spin filters." *Proteomics* 5: 1742-1745.

Martinetz, T.M. and Schulten, K.J. (1991). A neural-gas network learns topologies. *Artificial Neural Networks*. T. Kohonen, K. Mäkisara, O. Simula and J. Kangas. Amsterdam, North-Holland: 397-402.

Mortensen, P., Gouw, J.W., et al. (2009). "MSQuant, an Open Source Platform for Mass Spectrometry-Based Quantitative Proteomics." *Journal of Proteome Research:* 10.1021/pr900721e (only published online until yet).

Moxon, J.V., Padula, M.P., et al. (2009). "Challenges, Current Status and Future Perspectives of Proteomics in Improving Understanding, Diagnosis and Treatment of Vascular Disease " *Eur J Vasc Endovasc Surg* 38: 346-355.

Nesvizhskii, A.I., Keller, A., et al. (2003). "A Statistical Model for Identifying Proteins by Tandem Mass Spectrometry." *Anal. Chem.* 75: 4646-4658.

Oeljeklaus, S., Meyer, H.E., et al. (2009). "Advancements in plant proteomics using quantitative mass spectrometry." *Journal of Proteomics* 72(3): 545-554.

Old, W.M., Meyer-Arendt, K., et al. (2005). "Comparison of Label-free Methods for Quantifying Human Proteins by Shotgun Proteomics." *Molecular & Cellular Proteomics* 4: 1487-1502.

Park, S.K., Venable, J.D., et al. (2008). "A quantitative analysis software tool for mass spectrometry-based proteomics." *Nature Methods* 5(4): 319-322.

Picotti, P., Aebersold, R., et al. (2007). "The implications of proteolytic beckground for shotgun proteomics." *Molecular and Cellular Proteomics.*

Picotti, P., Bodenmiller, B., et al. (2009). "Full Dynamic Range Proteome Analysis of S. cerevisiae by Targeted Proteomics." *Cell* 138: 795-806.

Prince, J.T. and Marcotte, E.M. (2006). "Chromatographic alignment of ESI-LC-MS proteomics data sets by ordered bijective interpolated warping." *Anal. Chem* 78: 6140-6152.

Rappsilber, J., Mann, M., et al. (2007). "Protocol for micro-purification, enrichment, pre-fractionation and storage of peptides for proteomics using StageTips " *Nature Protocols* 2(8): 1896-1906.

Roewer, C., Vissers, J.P.C., et al. (2009). "Towards a proteome signature for invasive ductal breast carcinoma derived from label-free nanoscale LC-MS protein expression profiling of tumorous and glandular tissue." *Analytical Bioanalytical Chemistry* 395(8): 2443-2456.

Scheurer, S.B., Rybak, J.-N., et al. (2005). "Identification and relative quantification of membrane proteins by surface biotinylation and two-dimensional peptide mapping." *Proteomics* 5: 2718-2728.

Schneider, M., Bairoch, A., et al. (2005). "Plant Protein Annotation in the UniProt Knowledgebase." *Plant Physiology* 138: 59-66.

Schulte, D., Close, T.J., et al. (2009). "The International Barley Sequencing Consortium--At the Threshold of Efficient Access to the Barley Genome." *Plant Physiol.* 149(1): 142-147.

Seiffert, U., Jain, L.C., et al. (2005). *Bioinformatics using Computational Intelligence Paradigms*. Heidelberg, Springer.

Selbach, M. and Mann, M. (2006). "Protein interaction screening by quantitative immunoprecipitation combined with knock-down (QUICK)." *Nature Methods* 3: 981-983.

Sharma, K., Weber, C., et al. (2009). "Proteomics strategy for quantitative protein interaction profiling in cell extracts." *Nature Methods* 6(10): 741-744.

Shen (2005). "Automated 20 kpsi RPLC-MS and MS/MS with Chromatographic Peak Capacities of 1000-1500 and Capabilities in Proteomics and Metabolomics." *Anal. Chem.* 77: 3090-3100.

Silva, J.C., Denny, R., et al. (2005). "Quantitative Proteomic Analysis by Accurate Mass Retention Time Pairs." *Analytical Chemistry*_77: 2187-2200.

Silva, J.C., Gorenstein, M.V., et al. (2006). "Absolute quantification of proteins by LCMS^{E} - A virtue of parallel MS acquisition." *Molecular & Cellular Proteomics* 5: 144-156.

Sun, Q., Zybailov, B., et al. (2008). "PPDB, the Plant Proteomics Database at Cornell." *Nucleic Acids Research* 37: D969-D974.

Taylor, C.F., Paton, N.W., et al. (2007). "The minimum information about a proteomics experiment (MIAPE)." *Nature Biotechnology* 8: 887-893.

Thelen, J.J. and Peck, S.C. (2007). "Quantitative Proteomics in Plants: Choices in Abundance." *The Plant Cell* 19: 3339-3346.

Usadel, B., Nagel, A., et al. (2005). "Extension of the visualization tool MapMan to allow statistical analysis of arrays, display of corresponding genes, and comparison with known responses." *Plant Physiology* 138: 1195-204.

Usadel, B., Poree, F., et al. (2009). "A guide to using MapMan to visualize and compare Omics data in plants: a case study in the crop species, Maize." *Plant Cell Environment* 32: 1211-1229.

Vissers, J.P.C., Langridge, J.I., et al. (2007). "Analysis and quantification of diagnostic serum markers and protein signatures for Gaucher disease." *Molecular & Cellular Proteomics* 6: 755-766.

Vissers, J.P.C., Pons, S., et al. (2009). "The use of proteome similarity for the qualitative and quantitative profiling of reperfused myocardium." *Journal of Chromatography B* 877: 1317-1326.

Vogel, C. and Marcotte, E.M. (2008). "Calculating absolute and relative protein abundance from mass spectrometry-based protein expression data." *Nature Protocols* 3(9): 1444-1451.

Wang, H., Qian, W.-J., et al. (2005). "Development and Evaluation of a Micro- and Nano-Scale Proteomic Sample Preparation Method." *Journal of Proteome Research* 4(6): 2397-2403.

Wang, W., Zhou, H., et al. (2003). "Quantification of proteins and metabolites by mass spectrometry without isotopic labeling or spiked standards." *Analytical Chemistry* 75: 4818-4826.

Wang, X., Stoll, D.R., et al. (2006). "Peak Capacity Optimization of Peptide Separations in Reversed-Phase Gradient Elution Chromatography: Fixed Column Format." *Anal. Chem* 78: 3406-3416.

Washburn, M.P., Ulaszek, R., et al. (2002). "Analysis of Quantitative Proteomic Data Generated via Multidimensional Protein Identification Technology." *Anal. Chem.* 74(7): 1650-1657.

Weckwerth, W. (2008). "Integration of metabolomics and proteomics in molecular plant physiology – coping with the complexity by data-dimensionality reduction." *Physiol. Plant.* 132(2): 176-189.

Wessels, H.J.C.T., Vogel, R.O., et al. (2009). "LC-MS/MS as an alternative for SDS-PAGE in blue native analysis of protein complexes." *Proteomics* 9: 4221-4228.

Wienkoop, S., Morgenthal, K., et al. (2008). "Integration of metabolomic and proteomic phenotypes - Analysis of data-covariance dissects starch and RFO metabolism from low and high temperature compensation response in Arabidopsis thaliana." *Molecular & Cellular Proteomics* 7: 1725-1736.

Wilm, M. (2009). "Quantitative proteomics in biological research." *Proteomics* 9: 4590-4605.

Wilson, I.D., Nicholson, J.K., et al. (2005). "High Resolution "Ultra Performance" Liquid Chromatography Coupled to oa-TOF Mass Spectrometry as a Tool for Differential Metabolic Pathway Profiling in Functional Genomic Studies." *Journal of Proteome Research* 4: 591-598.

Winter, D., Pipkorn, R., et al. (2009). "Separation of peptide isomers and conformers by ultra performance liquid chromatography." *Journal of Separation Science* 32(8): 1111-1119.

Wisniewski, J.R., Zougman, A., et al. (2009). "Universal sample preparation method for proteome analysis." *Nat Meth* 6(5): 359-362.

Witzel, K., Surabhi, G.-K., et al. (2007). "Quantitative Proteome Analysis of Barley Seeds Using Ruthenium(II)-tris-(bathophenanthroline-disulphonate) Staining." *J. Proteome Res.* 6(4): 1325-1333.

Wolters, D.A., Washburn, M.P., et al. (2001). "An Automated Multidimensional Protein Identification Technology for Shotgun Proteomics." *Anal. Chem.* 73(23): 5683-5690.

Wong, J.W.H., Sullivan, M.J., et al. (2007). "Computational methods for the comparative quantification of proteins in label-free LC^n-MS experiments." *Briefings in bioinformatics* 9(2): 156-165.

Wu, C.H., Huang, H., et al. (2004). "The iProClass integrated database for protein functional analysis." *Computational Biology and Chemistry* 28: 87-96.

Yeung, Y.-G., Nieves, E., et al. (2008). "Removal of detergents from protein digests for mass spectrometry analysis." *Analytical Biochemistry* 382(2): 135-137.

Yocum, A.K. and Chinnaiyan, A.M. (2009). "Current affairs in quantitative targeted proteomics: multiple reaction monitoring-mass spectrometry." *Briefings in Functional Genomics and Proteomics* 8(2): 145-157.

Zhang, B., VerBerkmoes, N.C., et al. (2006). "Detecting differential and correlated protein expression in label-free shotgun proteomics." *Journal of Proteome Research* 5(11): 2909-2918.

In: Proteomics
Editor: Giselle C. Rancourt

ISBN: 978-1-61668-691-8

Chapter 7

INSIGHTS FROM PROTEOMICS INTO MILD COGNITIVE IMPAIRMENT, LIKELY THE EARLIEST STAGE OF ALZHEIMER'S DISEASE

Renã A. Sowell and D. Allan Butterfield
University of Kentucky, Lexington, KY, USA

ABSTRACT

Mild cognitive impairment (MCI) is arguably the earliest form of Alzheimer's disease (AD). Better understanding of brain changes in MCI may lead to the identification of therapeutic targets to slow the progression of AD. Oxidative stress has been implicated as a mechanism associated with the pathogenesis of both MCI and AD. In particular, among other markers, there is evidence for an increase in the levels of protein oxidation and lipid peroxidation in the brains of subjects with MCI. Several proteins are oxidatively modified in MCI brain, and as a result individual protein dysfunction may be directly linked to these modifications (e.g., carbonylation, nitration, modification by HNE) and may be involved in MCI pathogenesis. Additionally, Concanavalin-A-mediated separation of brain proteins has recently led to the identification of key proteins in MCI and AD using proteomics methods. This chapter will summarize important findings from proteomics studies of MCI, which have provided insights into this cognitive disorder and have led to further understanding of potential mechanisms involved in the progression of AD.

Keywords: Mild Cognitive Impairment, proteomics, oxidative modifications, Alzheimer's disease

1. Introduction

Mild cognitive impairment (MCI) can be considered as the earliest form of Alzheimer's disease (AD) existing as a transitional state between normal aging and AD [1-3]. MCI exists in two forms: amnestic MCI and nonamnestic MCI [2, 3]. Amnestic MCI patients are able to perform normal daily living activities and have no signs of dementia; however, they do have cognitive complaints that include bursts of episodic memory loss [1, 4]. In some cases, amnestic MCI patients can develop AD at a rate of ~10 to 15% annually, however in other cases, the patients revert back to normal conditions [5]. Pathologic characteristics of MCI are similar to those of AD. For example, MCI patients have hippocampal, entorhinal cortex (EC), and temporal lobe atrophy based on magnetic resonance imaging studies [6-8], synapse loss, neuronal loss, low cerebrospinal fluid (CSF)-resident β amyloid levels [6], genetic risk factors including preponderance in APOE4 allele [9, 10], and increased levels of oxidative stress [11-20].

Oxidative stress is one of the underlying indices associated with MCI, AD, and other neurodegenerative disorders such as Parkinson's disease and amyotrophic lateral sclerosis. Specifically in MCI, there is substantial evidence for increased levels of oxidative stress in the brains and in plasma of MCI subjects [11-23]. Our laboratory has reported an increase in the levels of protein carbonyls (PCO) [11, 16] and 3-nitrotyrosine (3NT)-modified proteins [21], both of which are markers of protein oxidation. Additionally, we have reported an increase in the levels of 4-hydroxynonenal-(HNE) bound proteins, indicating an increase in the levels of lipid peroxidation products [13]. Others have observed decreases in the levels of antioxidant enzymes and antioxidant enzymatic activity in brain and in plasma [22-24], increased levels of oxidative stress in nuclear and mitochondrial DNA [25, 26], increased levels of isoprostanes [27], and increased lipid peroxidation as measured by free HNE levels, thiobarbituic substances, and malondialdehyde [16, 20]. It is believed that oxidative stress also is related to several vascular factors, such as heart disease, hypertension, and diabetes mellitus that conceivably contribute to the conversion of MCI into AD.

It is important to understand more about the events that lead to the progression of AD from MCI in order to develop potential therapeutics that can delay or stop AD onset. Thus, proteomics can provide considerable insight into specific pathways that are influenced by MCI and which eventually aid in the progression of disease. To this end, we and others have investigated the changes associated with the proteomes of MCI subjects relative to normal age-matched healthy controls [11, 19, 28-33]. These studies include the search for candidate biomarkers of MCI which eventually lead to AD [29, 30, 33], changes in the expression levels of proteins [28], specific levels of protein oxidation as measured by PCO [11], 3NT-modified proteins [19], and lipid peroxidation as measured by HNE-bound proteins [32]. More recently, we have also investigated other post-translational modifications that change in subjects with MCI such as glycosylation [31]. This chapter summarizes the key findings from proteomics and redox proteomics studies in MCI and their implications in AD research.

2. Two-Dimensional (2D) Gel Electrophoresis (GE) Based Proteomics

The proteomics techniques used in the studies described herein follow the general approach outlined in Figure 1. Here proteins are extracted from brain, CSF, plasma, or other bodily tissues obtained from MCI subjects and normal age-matched controls. Extracted proteins are subjected to isoelectric focusing (IEF)/sodium dodecyl sulfate (SDS) polyacrylamide gel electrophoresis (PAGE), better known as 2DGE. In this approach, proteins are separated in a first dimension based on their isoelectric point and in a second dimension based on their migration rate through the gel, which often corresponds to molecular weight. Image analysis software is used to align spots across the gels obtained from all of the samples, and protein spots that exhibit significant changes (based on Student's t-tests or analysis of variance) in expression levels between MCI and controls are excised. Excised spots are subjected to in-gel trypsin digestion and analyzed using matrix assisted laser desorption ionization (MALDI) or electrospray ionization (ESI) mass spectrometry (MS). Data from MS experiments are then submitted to appropriate databases using search engines such as MASCOT [34] for protein identification.

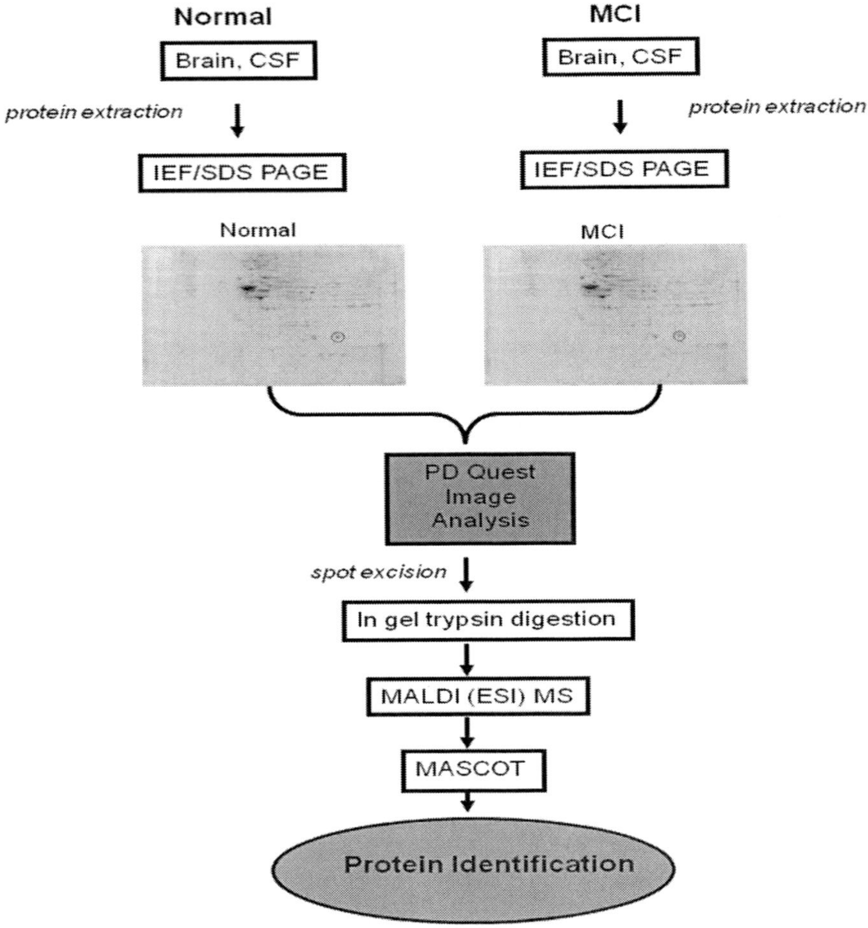

Figure 1. Schematic overview of the 2D GE experiment on brain or CSF from subjects with MCI and age-matched controls.

This general approach can also be adapted for the analysis of post-translational modifications. For example, changes in glycosylation of proteins can be analyzed by using affinity chromatography for purification of glycoproteins. Extracted proteins can be separated with Concanavalin-A lectin affinity columns which isolate proteins that contain asparagine (N)-linked carbohydrates. In some cases, Con-A may also have nonspecific interactions and isolate proteins based on its hydrophobic binding domain [35].

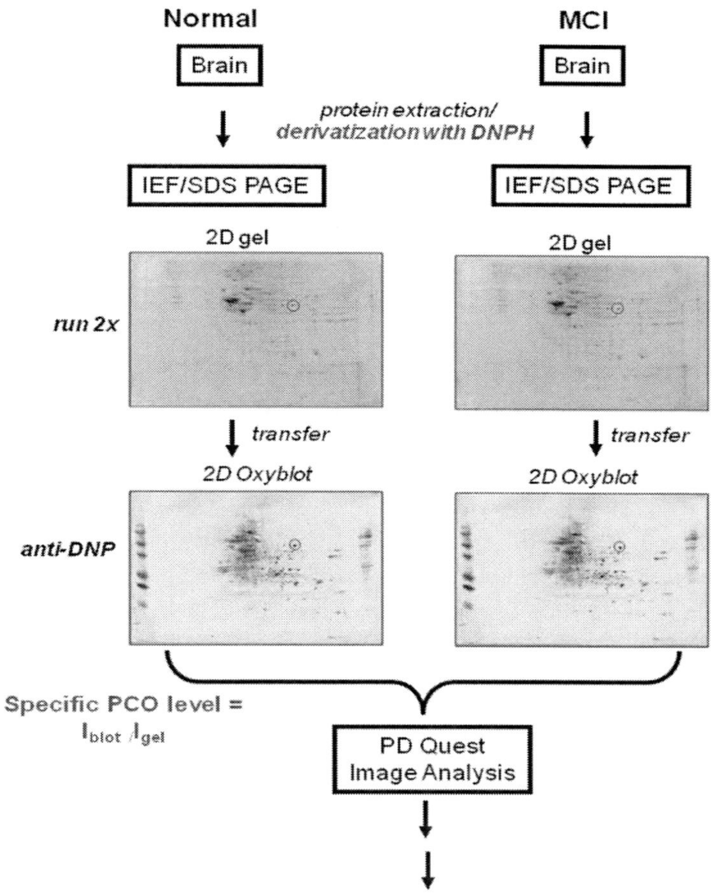

Figure 2. Schematic overview of the redox proteomics approach applied for the analysis of oxidatively modified proteins such as protein carbonyls (PCO), 3-nitrotyrosine (3NT) modified proteins and HNE-modified proteins. Extracted proteins are derivatized with 2,4-dinitrophenylhydrazine (DNPH) only for the analysis of PCO and separated with IEF/SDS PAGE. 2D gels are then transferred onto a nitrocellulose membrane and 2D Oxyblots are probed with anti-DNP (or anti-3NT, anti-HNE) antibodies and visualized using a secondary antibody linked with a colorimetric alkaline phosphatase assay. Specific oxidative levels of proteins are calculated by normalizing the intensity of spots in the 2D Oxyblot (I_{blot}) to the intensity of the corresponding spot in a 2D gel (I_{gel}). This calculation is carried out similarly for PCO, 3NT, and HNE. Protein spots exhibiting significant changes in oxidative modification are then excised, digested in-gel by trypsin, analyzed with MALDI or ESI-MS/MS, and identified with database searching as illustrated in Figure 1.

Oxidative modification of proteins can also be detected using the 2D GE approach with the incorporation of Western blotting analysis [36]. Figure 2 shows a schematic of the general approach used to detect PCO, 3NT-modified proteins, and HNE-bound proteins. As shown in Figure 2, for the detection of PCO, extracted proteins are derivatized with 2,4-dinitrophenylhydrazine (DNPH), which forms a Schiff base with carbonyl groups on proteins. DNPH-derivatized proteins are then separated using 2D GE, and the spots on gels are transferred onto a nitrocellulose membrane forming a 2D Western blot or 2D Oxyblot. Immunochemical detection using a primary anti-DNP antibody that recognizes DNP hydrazone adducts is applied to the 2D Oxyblots, and oxidized spots are visualized with a secondary antibody linked to a colorimetric alkaline phosphatase assay. Similarly, for the detection of 3NT- and HNE-modified proteins non-derivatized extracted proteins are separated with 2D GE, transferred onto an 2D Oxyblot, and immunochemically detected with anti-3NT and anti-HNE, antibodies, respectively. Imaging analysis software and statistical approaches are applied as illustrated in Figure 1 in order to align 2D images and identify spots that have significant changes in oxidative modification. Specific carbonyl (or 3NT, HNE) levels of proteins are measured by normalizing the density of spots in the 2D Oxyblot, to the density of the same spot in a 2D gel analysis of the sample (separate experiment). Protein spots of interest (those exhibiting significant elevation or reduction in oxidative modification) are then excised from the gel, tryptically digested, analyzed by MS, and identified as described in Figure 1.

3. Candidate Biomarkers in CSF for the Progression of MCI to AD

CSF presents another biological fluid that can relay specific information about neurological molecular changes because the fluid encompasses the extracellular space surrounding the brain. Table 1 lists proteins that were identified as having significant changes in expression in CSF of MCI subjects relative to normal controls. Kim and coworkers have performed proteomics analysis on CSF from normal cognitive controls, MCI, and AD patients and identified three proteins which may be candidate markers for the diagnosis MCI and its progression into AD[29, 30]. The protein, fibrinogen γ-A chain, was detected as having a gradual elevation of expression in MCI and AD [30]. Fibrinogen γ -A chain is a part of the 340 kDa hexameric soluble glycoprotein (consisting of α, β, and γ chains) that is synthesized in the liver. This protein is involved in the polymerization of fibrin, blood coagulation, signal transduction, platelet activation and binding, and thrombin binding [37]. Fibrinogen has also been shown to have elevated expression levels during inflammation and in cardiovascular disease [37]; thus, its elevation in MCI may be reflective of early events of neuroinflammation. The other two proteins, plasma retinol-binding protein (RBP) and haptoglobin precursor allele 1, were detected as having a significant decrease in expression in CSF from MCI (and AD) patients relative to normal age-matched controls by 38% and 63%, respectively [29]. RBP is a 21 kDa carrier protein that tightly binds retinol allowing for it to freely circulate through plasma. Haptoglobin is a tetrameric protein that is a part of the acute phase response and binds free hemoglobin. Through binding of hemoglobin, haptoglobin inhibits oxidative activity of hemoglobin, prevents iron loss in the kidneys, and protects the kidneys against damage that could be caused by hemoglobin [38]. The effects of decreased expression of RBP and haptoglobin in MCI is not clearly understood [29].

Table 1. Candidate Biomarker Proteins in CSF of MCI

Protein	Change in MCI	Ref
C3a des-Arg	↑	Simonsen *et al.* 2007
C4a des-Arg	↑	Simonsen *et al.* 2007
Fibrinogen γ-chain A	↑	Lee *et al.* 2007
Haptoglobin precursor allele 1	↓	Jung *et al.* 2008
Phosphorylated osteopontin C-terminal fragment	↑	Simonsen *et al.* 2007
Plasma retinol-binding protein	↓	Jung *et al.* 2008
Ubiquitin	↑	Simonsen *et al.* 2007
β_2-Microglobulin	↑	Simonsen *et al.* 2007

Using 2D GE coupled to surface-enhanced laser desorption ionization (SELDI)-MS, Simonsen *et al.* identified a panel of 17 proteins that may be potential biomarkers of patients with MCI that convert to AD and of patients with MCI who do not progress to AD [33]. Of the 17 proteins, four proteins were down-regulated and 13 proteins were up-regulated in CSF of MCI patients that converted to AD relative to stable MCI patients and normal healthy controls [33]. Five proteins were identified with MS and have elevated expression in MCI patients that progress to AD: ubiquitin, C3a anaphylatoxin des-Arg, C4a anaphylatoxin des-Arg, β2-Microglobulin, and phosphorylated osteopontin C-terminal fragment. β-amyloid can bind to C1q and subsequently activate the complement cascade resulting in the production of C3a and C4a, as well as C5a peptides [39]. Osteopontin is a glycoprotein and proinflammatory cytokine involved in bone synthesis and various aspects of immunity such as chemotaxis [40], cell adhesion and wound healing [41], cell activation and cytokine production [41], and apoptosis [42, 43]. Elevation of the complement peptides and osteopontin in MCI patients that progress to AD suggests that innate immunity including inflammation in MCI patients may become activated and stay activated in the progression of disease. β2-Microglobulin is a part of the class I major histocompatibility complex and mediates amyloid fibril formation *in vitro* [44]and in the presence of transition metal cations [45]. Ubiquitin is used to target proteins for degradation by the 26S proteasome [46], and has been immunhistochemically shown to be present in neurofibrillary tangles (NFT) and senile plaques (SP) [47].

4. Proteomics Analyses of Brain from MCI Patients

An alternative approach to 2D GE that was recently used in proteomic comparisons of brain from MCI subjects relative to normal cognitive controls, is the PowerBlot proteomic approach (BD Transduction Laboratories). This approach uses a large-scale Western Blot approach to identify 750+ proteins simultaneously in a single experiment. Using the PowerBlot approach, Ho *et al.* detected 50 candidate proteins that had >2.0 fold-change in the EC region of MCI patients relative to normal cognitive controls [28]. Of the 50 proteins detected, 23 proteins were identified and could be functionally clustered into the following categories: neurotransmitter-related, cytoskeleton/cell adhesion, cell cycle/cell proliferation related, apopotosis related, transcription/translation related, and others. Neurotransmitter-related, apoptosis-related, and transcription/translation-related proteins were decreased in the EC of MCI patients, while cytoskeleton and cell cycle-related proteins included both increased and decreased proteins in MCI patients [28]. Several of these functional categories are similar to those observed in oxidatively modified proteins in MCI hippocampal brain regions and are discussed below.

5. Redox Proteomics Analyses of Brain from MCI Patients

Table 2 lists significantly elevated oxidatively modified proteins (i.e., PCO-, 3NT-, and HNE-modified) from the hippocampal and inferior parietal lobule (IPL) brain regions of MCI subjects relative to age-matched normal controls that were identified by our laboratory.

Table 2. Functional Categorization of Oxidatively Modified Proteins Identified in Brains of MCI Patients

Functional Categories	Protein	Oxidative Modifications
Energy/mitochondrial dysfunction	α-enolase	**PCO, 3NT, HNE**
	aldolase	**3NT**
	malate dehydrogenase	**3NT**
	glucose-regulated protein precursor	**3NT**
	lactate dehydrogenase	**HNE**
	phosphoglycerate kinase	**HNE**
	pyruvate kinase	**PCO, HNE**
	ATP synthase α-chain	**HNE**
Lipid abnormalities & cholinergic dysfunction	neuropolypeptide h3	**HNE**
Excitotoxicity	glutamine synthetase	**PCO**
Cell cycle, tau phosphorylation, A β production	PIN1	**PCO**
Neuritic abnormalities & structural dysfunction	DRP2	**3NT**
	fascin 1	**3NT**
	β actin	**HNE**
Antioxidant defense/Detoxification system dysfunction	GSTM3	**3NT**
	MRP3	**3NT**
	peroxiredoxin 6	**3NT**
	HSP70	**3NT, HNE**
	carbonyl reductase 1	**HNE**
Cell signaling dysfunction	14-3-3-γ	**3NT**
Protein synthesis alterations	Initiation factor α	**HNE**
	Elongation factor Tu	**HNE**

To-date these are the only reports of specific oxidatively modified proteins in MCI brain that may be relevant to the progression of AD [11, 19, 32]. These proteins can be grouped into several functional categories and were significantly oxidatively modified by one or more of the three oxidative parameters: PCO, 3NT, and HNE.

5.1. Energy or Mitochondrial Dysfunctions and Alterations

As listed in Table 2, several proteins involved in energy and/or mitochondrial-related pathways have significantly elevated levels of oxidative modification in the hippocampal or IPL regions of brains from subjects with MCI. These proteins are α-enolase, aldolase, pyruvate kinase (PK), malate dehydrogenase (MDH), lactate dehydrogenase (LDH), ATP synthase, phosphoglycerate kinase (PGK1), and glucose regulated protein precursor. Glycolysis plays an important role in supplying energy to the brain because glucose is the primary source of energy. Alterations in glucose metabolism and tolerance have been identified in brains of MCI and AD patients from positron emission tomography scanning, [48-50] and oxidatively modified glycolytic proteins have been identified in MCI and AD brain, and models thereof [12, 51]. α-Enolase, aldolase, PK, PGK1, and LDH are enzymes involved in or related to the glycolysis pathway. Increased oxidation of α-enolase, LDH, and PK has been shown to lead to loss of protein function measured by decreased enzymatic activity in MCI brain [11, 32]. Alterations in glycolysis could lead to less ATP production which is detrimental to cells requiring ATP to carry out normal functions, including signal transduction, maintenance of ion gradients, and protein synthesis, and detrimental to ATPases which are responsible for proper maintenance of ion pumps, lipid asymmetry, and intracellular communication. The observed impairments to glycolytic proteins in MCI brain suggest that energy metabolism is a key player in the progression of MCI to AD. This is further supported by the increased oxidation in MDH, ATP synthase α-chain, and glucose regulated protein precursor. ATP synthase α-chain is a component of complex V which plays a key role in energy production and undergoes a series of coordinated conformational changes in order to produce ATP. Oxidative modification to ATP synthase leads to reduced enzymatic activity [32]. Because ATP synthase is involved in the electron transport chain (ETC), alterations to its activity could result in an electron leakage from ETC carrier molecules, which would lead to an increase in reactive oxygen species (ROS). These ROS could then contribute to the observed increase in oxidative stress parameters in MCI brain [11, 13, 16, 20, 21, 25-27]. An overall decrease in ATP production due to dysfunction of glycolytic enzymes, ATP synthase, and glucose regulated protein precursor (from oxidative modification) could ultimately lead to Ca2+ dyshomeostasis and make neurons susceptible to excitotoxicity and cell death. From these studies it is apparent that potential preventative targets for AD could be targeted to restoring energy metabolism in earlier disease stages in MCI. In contrast to the usual observation of decreased enzymatic activity of oxidatively modified proteins, oxidative modification of MDH leads to increased activity [32]. The basis of this unusual observation remains unclear.

5.2. Neuritic Abnormalities/Structural Dysfunction

Oxidatively modified proteins in MCI related with neuritic and structural functions are dihydropyrimidinase like-2 (DRP2), β-actin, and fascin 1. DRP-2 is involved in axonal outgrowth and transmission and modulation of extracellular signals through the protein collapsin [52, 53]. In AD, DRP-2 also has increased oxidation [54, 55], which may be reflective of increasing neuritic degeneration, shortened dendritic length, and synapse loss as MCI progresses to AD. β-actin is crucial for proper maintenance of cellular and cytoskeletal integrity and morphology. High levels of actin can be found in growth cones, presynaptic terminals and dendritic spines, and thus its oxidation could lead to elongation of growth cones and synapse loss. Alterations in cellular integrity could be detrimental to cellular trafficking of key proteins involved in neurotransmission. Fascin 1, also known as p55, is a structural protein involved in cell adhesion and motility [56-58] and is used as a marker of normal dendritic function [59]. Overall, oxidation of structural proteins which could result in altered functionality ultimately can lead to impaired structural integrity, shortened dendritic lengths and faulty axonal growth, loss of interneuronal connections and poor neurotransmission. Neuritic abnormalities and structural dysfunction are documented in AD brain and thus may be key events in the loss of neurotransmission in MCI to AD.

5.3. Excitotoxicity

Overstimulation of neurons can result from an increase in the levels of extracellular glutamate. Glutamine synthetase, which converts glutamate to glutamine, was oxidatively modified in MCI brain and has been shown to have decreased activity [11]. Thus, decreased glutamine synthetase activity directly leads to a buildup in glutamate, which can lead to excitotoxicity. This phenomenon also affects Ca^{2+} homeostasis and eventually leads to neuronal death. Similar changes in glutamine synthetase oxidation and activity were observed in AD brain [60-62], and suggest that synapse loss observed in AD brain occurs early on in MCI with a contribution by excitotoxicity.

5.4. Lipid Abnormalities and Cholinergic Dysfunction

Neuropolypeptide h3 [(also known as phosphatidylethanolamine binding protein (PEBP)] is an enzyme involved in acetylcholine production and may play roles in phospholipid asymmetry. Oxidation of neuropolypeptide h3 in MCI brain and possible loss of function correlates well with the already known cholinergic loss observed in AD brain [63-66]. Also, loss of phospholipid asymmetry has been reported in MCI and AD brain [67-69], and thus oxidation of PEBP could play a role in lipid peroxidation events which lead to cellular apoptosis.

5.5. Antioxidant Defense/Detoxification System Dysfunction

Proteins involved in the antioxidant defense and detoxification system work to remove harmful species such as free radicals and toxic compounds from the cell. Peroxiredoxin 6 (PR6), multidrug resistance protein 3 (MRP3), glutathione-S-transferase Mu 3 (GSTM3), heat shock protein 70 (HSP70), and carbonyl reductase are oxidatively modified brain proteins in MCI. PR6 is an antioxidant enzyme that reduces the reactive nitrogen species, peroxynitrite, and was detected as having elevated nitration levels in MCI. PR6 also reduces reactive phospholipids and hydroperoxides [70] and has other roles which include cell differentiation, apoptosis, and detoxification [71]. Nitration of PR6, which could lead to loss of function, may result in increased levels of nitrated proteins, such as those detected in MCI brain [19, 21] . PR6 and GSTM3, a detoxification enzyme, form a complex that alters individual enzymatic activities [71] but which works to protect cells from toxic species such as HNE. GST catalyzes the conjugation of the low molecular weight intracellular thiol, glutathione, with toxins, and these toxins are transported out of the cell by MRP [72-74]. Oxidation and potential loss of function of PR6, MRP3, and GSTM3 could impair the cell's ability to remove toxicants leading to an increase in toxic species that subsequently attack cellular molecules (e.g., increased PCO, 3NT, or HNE) and lead to cell death. These observations in

MCI brain are consistent with changes to MRP, GST, and PR6 which are observed in AD brain [73, 75], and demonstrate that proper antioxidant and detoxification defense systems may help to delay the progression of MCI to AD.

HSP70 is a molecular chaperone protein that repairs misfolded proteins and helps in the transportation of misfolded proteins to the proteasome. HSP70 belongs to the class of HSPs that also protect proteins from various stresses, such as oxidative damage [76]. Nitration of HSP70, leading to loss of function, could result in buildup of misfolded proteins and hence protein aggregates and "clogging" of the proteasome. Carbonyl reductase is an enzyme that reduces carbonyl compounds to their corresponding alcohols. HNE-modification of carbonyl reductase is rather interesting considering that it has been shown to reduce HNE levels [77], and thus its oxidative modification by HNE would lead to an increase in HNE available for attack on proteins such as those HNE-modified proteins identified in subjects with MCI [32].

5.6. Cell Signaling Dysfunction

14-3-3 γ belongs to a family of scaffolding proteins that normally bridges glycogen synthase kinase 3β (GSK-3β) and tau by forming a multiprotein tau phosphorylation complex [78]. Other functions include signal transduction, protein trafficking, and metabolism [79, 80]. 14-3-3 γ was observed as nitrated in MCI brain and has previously been observed to have elevated expression levels in AD brain [81, 82] and in AD CSF [83]. Nitration of 14-3-3 γ may contribute to tau hyperphosphorylation and thus NFT formation and dysfunction in cell signaling, events which are consistent with changes observed in AD brain.

5.7. Cell Cycle; Tau Phosphorylation; Aβ Production

Peptidyl-prolyl *cis/trans* isomerase 1 (Pin1) is a multifunctional protein involved in the cell cycle, tau phosphorylation, and Aβ production and regulates cellular processes such as protein folding, transcription, intracellular transport, and apoptosis [84-86]. Pin1 is oxidized in MCI brain [11] and has been previously reported as oxidized in AD brain [55]. Pin1, through its interactions with kinases and phosphatases such as GSK-3β, can directly regulate the phosphorylation of the tau protein [84, 87]. Thus, inactivation of

Pin1 as a result of oxidative modification directly leads to hyperphosphorylation of tau and an increase in NFT. Pin1 has also been shown to bind to APP and influence the production of Aβ [84, 88]. Altered regulation of cell cycle processes by oxidized Pin1 may be related to elevation of cell cycle proteins in brain of MCI subjects [89]. Therefore, alterations to Pin1 activity also may lead to an increase in Aβ and SP. Oxidative impairment of Pin1 in early stages of AD, such as MCI, is consistent with and likely contributes to the major pathological hallmarks of AD: SP, NFT, and synapse loss.

5.8. Protein Synthesis Alterations

Initiation factor α (eIF-α) and elongation factor Tu (EF-Tu) are proteins involved in protein synthesis. eIF-α has been reported to have roles in cell proliferation and senescence [90], cytoskeletal organization [91], and apoptosis [92]. EF-Tu is a nuclear-encoded protein that assists in the translation of proteins in the mitochondria [93]. Specifically, EF-Tu binds aminoacylated tRNA in the cytoplasm and hydrolyzes GTP in order to allow the aminoacylated tRNA to enter the A site of the ribosome. Nitration of these proteins can directly influence protein synthesis. Decreased protein synthesis has been reported in MCI and AD [94-96], and thus these alterations are consistent with these reports. Alterations to protein synthesis in MCI may result in a reduction of key proteins necessary to combat many of the cellular insults observed in AD brain, which not only could result in compromised neuronal functions, but also contribute to progression of MCI to AD.

6. Concanavalin-A Associated Glycoproteins in Brain Regions from MCI Patients

DRP-2, glucose-regulated protein 78 (GRP78), protein phosphatase-related protein Sds-22, glial fibrillary acidic protein (GFAP), and β-synuclein were isolated in the ConA associated protein fraction using lectin affinity chromatography coupled to 2D GE and identified as having altered levels in the brains of subjects with MCI relative to age-matched controls [31]. DRP-2 and GRP78 were detected as significantly decreased and GFAP and protein phosphatase –related protein Sds22 as significantly increased in the hippocampal brain region of MCI patients, while β-synuclein was significantly decreased in the inferior parietal lobule region of MCI patients relative to age-matched controls [31]. DRP-2 as (discussed above) is a structural protein involved in axonal outgrowth and neuronal communication thus, its decreased expression may be indicative of neuritic dysfunction that occurs early in MCI and continues with disease progression into AD. GRP78 is an endoplasmic reticulum (ER) associated protein that belongs to the HSP70 protein family and is involved with the unfolfed protein response (UPR). HSP70 is significantly oxidatively modified in MCI brain (see Table 2). Because GRP78 normally reduces the levels of amyloid precursor protein (APP) and Aβ40 and 42 secretion [97], decreased expression of GRP78 in MCI brain could possibly play roles in the elevation of APP and Aβ levels found in AD brain. Also, alteration to GRP78 expression may disrupt Ca2+ homeostasis [98]. Conflicting reports of GRP78 expression in AD have been reported [99, 100], and thus its role in MCI progression to AD is not completely clear. Activation of the UPR in the ER might also mean that GRP78 is less available for refolding other damaged proteins or shuttling them to the 26S proteasome for degradation.

Protein phosphatase-related protein Sds22 is involved in the cell cycle and was detected as increased in MCI brain, the significance of this increase in the progression of MCI to AD is yet to be determined but as noted above, cell cycle proteins are elevated in brains of subjects with MCI [89]. β-synuclein is involved in synaptic functions, similar to the functions of α-synuclein in Parkinson's disease. β-synuclein also binds to Aβ [101] and has previously been shown by our laboratory to be oxidized *in vivo* following injection of Aβ(1-42) [102[. Decreased expression of β-synuclein in MCI brain could be related with altered synaptic functions which occur also due to the oxidatively modified proteins involved in synaptic functions mentioned above. GFAP, a glial specific intermediate filament protein is significantly increased in MCI brain. GFAP is involved in cytoskeletal integrity and maintenance of cellular shape and movement in astrocytes. Increased expression of GFAP in MCI brain is consistent with increased expression levels in AD [103] and with inflammation related to NFT and SP [104]. This finding provides further evidence supporting the notion that neuroinflammation is an event that occurs in the early stages of AD (MCI) and continues with disease progression.

7. Conclusion

This chapter has summarized some of the key findings from proteomics studies involving comparisons of brain and CSF in MCI subjects relative to normal age-matched controls. Candidate biomarkers of MCI that may help in early AD diagnosis were identified in CSF and may be useful as additional markers for AD diagnosis to the traditional tau (τ and p), and Aβ40 and 42 markers. Expression and redox proteomics analyses of various brain regions (e.g., EC, IPL, and hippocampus) revealed that a number of processes are altered in MCI including, neurotransmitter-related, apoptosis-related, energy/mitochondrial dysfunction, neuritic abnormalities/structural dysfunction, excitotoxicity, lipid abnormalities and cholinergic dysfunction, antioxidant defense/detoxification systems, cell signaling dysfunction, cell cycle/tau phosphorylation/Aβ production, and transcription/translation (protein synthesis) alterations. It is apparent that MCI to AD progression is a multifactorial process in which many pathways may be potential targets for intervening therapeutics. A large number of energy-related proteins were oxidatively modified in MCI further supporting the concept that normal energy maintenance is crucial and lacking in MCI and AD brain. In addition to oxidative modifications, concanavalin-A associated proteins have altered expression levels in IPL and hippocampal regions of MCI patients. These proteins are involved in structural integrity and molecular chaperoning, and altered levels of these proteins are congruent with the observed oxidative modifications and hence alterations of several structural and antioxidant defense/detoxification proteins. Proteomics has provided much insight into pathways that are related to MCI and with its progression to AD. Each of these pathways should be further investigated for their potential as therapeutic targets for early AD diagnosis, treatment, and/or prevention.

Acknowledgments

This work was supported by NIH grants to D.A.B. [AG-05119; AG-10836; AG-029839]. R.A.S. is a UNCF-Merck postdoctoral scholar.

REFERENCES

[1] Morris, J. C., Mild cognitive impairment and preclinical Alzheimer's disease. *Geriatrics* 2005, *Suppl*, 9-14.

[2] Petersen, R. C., Mild cognitive impairment: transition between aging and Alzheimer's disease. *Neurologia* 2000, *15* (3), 93-101.

[3] Portet, F.; Ousset, P. J.; Touchon, J., [What is a mild cognitive impairment?]. *Rev. Prat.* 2005, *55* (17), 1891-4.

[4] Petersen, R. C.; Smith, G. E.; Waring, S. C.; Ivnik, R. J.; Tangalos, E. G.; Kokmen, E., Mild cognitive impairment: clinical characterization and outcome. *Arch Neurol.* 1999, *56* (3), 303-8.

[5] Petersen, R. C., Mild cognitive impairment clinical trials. *Nat. Rev. Drug Discov.* 2003, *2* (8), 646-53.

[6] Chertkow, H.; Bergman, H.; Schipper, H. M.; Gauthier, S.; Bouchard, R.; Fontaine, S.; Clarfield, A. M., Assessment of suspected dementia. *Can. J. Neurol. Sci.* 2001, *28 Suppl 1*, S28-41.

[7] Devanand, D. P.; Pradhaban, G.; Liu, X.; Khandji, A.; De Santi, S.; Segal, S.; Rusinek, H.; Pelton, G. H.; Honig, L. S.; Mayeux, R.; Stern, Y.; Tabert, M. H.; de Leon, M. J., Hippocampal and entorhinal atrophy in mild cognitive impairment: prediction of Alzheimer disease. *Neurology* 2007, *68* (11), 828-36.

[8] Du, A. T.; Schuff, N.; Amend, D.; Laakso, M. P.; Hsu, Y. Y.; Jagust, W. J.; Yaffe, K.; Kramer, J. H.; Reed, B.; Norman, D.; Chui, H. C.; Weiner, M. W., Magnetic resonance imaging of the entorhinal cortex and hippocampus in mild cognitive impairment and Alzheimer's disease. *J. Neurol. Neurosurg. Psychiatry* 2001, *71* (4), 441-7.

[9] Negash, S.; Petersen, L. E.; Geda, Y. E.; Knopman, D. S.; Boeve, B. F.; Smith, G. E.; Ivnik, R. J.; Howard, D. V.; Howard, J. H., Jr.; Petersen, R. C., Effects of ApoE genotype and mild cognitive impairment on implicit learning. *Neurobiol. Aging* 2007, *28* (6), 885-93.

[10] Ramakers, I. H.; Visser, P. J.; Aalten, P.; Bekers, O.; Sleegers, K.; van Broeckhoven, C. L.; Jolles, J.; Verhey, F. R., The association between APOE genotype and memory dysfunction in subjects with mild cognitive impairment is related to age and Alzheimer pathology. *Dement Geriatr Cogn. Disord.*2008, *26* (2), 101-8.

[11] Butterfield, D. A.; Poon, H. F.; St Clair, D.; Keller, J. N.; Pierce, W. M.; Klein, J. B.; Markesbery, W. R., Redox proteomics identification of oxidatively modified hippocampal proteins in mild cognitive

impairment: insights into the development of Alzheimer's disease. *Neurobiol. Dis.* 2006, *22* (2), 223-32.

[12] Butterfield, D. A.; Reed, T.; Newman, S. F.; Sultana, R., Roles of amyloid beta-peptide-associated oxidative stress and brain protein modifications in the pathogenesis of Alzheimer's disease and mild cognitive impairment. *Free Radic. Biol. Med.* 2007, *43* (5), 658-77.

[13] Butterfield, D. A.; Reed, T.; Perluigi, M.; De Marco, C.; Coccia, R.; Cini, C.; Sultana, R., Elevated protein-bound levels of the lipid peroxidation product, 4-hydroxy-2-nonenal, in brain from persons with mild cognitive impairment. *Neurosci. Lett.* 2006, *397* (3), 170-3.

[14] Cenini, G.; Sultana, R.; Memo, M.; Butterfield, D. A., Elevated levels of pro-apoptotic p53 and its oxidative modification by the lipid peroxidation product, HNE, in brain from subjects with amnestic mild cognitive impairment and Alzheimer's disease. *J. Cell Mol. Med.* 2008, *12* (3), 987-94.

[15] Ding, Q.; Markesbery, W. R.; Cecarini, V.; Keller, J. N., Decreased RNA, and increased RNA oxidation, in ribosomes from early Alzheimer's disease. *Neurochem. Res.* 2006, *31* (5), 705-10.

[16] Keller, J. N.; Schmitt, F. A.; Scheff, S. W.; Ding, Q.; Chen, Q.; Butterfield, D. A.; Markesbery, W. R., Evidence of increased oxidative damage in subjects with mild cognitive impairment. *Neurology* 2005, *64* (7), 1152-6.

[17] Lovell, M. A.; Markesbery, W. R., Oxidative damage in mild cognitive impairment and early Alzheimer's disease. *J. Neurosci. Res.* 2007, *85* (14), 3036-40.

[18] Murphy, M. P.; Beckett, T. L.; Ding, Q.; Patel, E.; Markesbery, W. R.; St Clair, D. K.; LeVine, H., 3rd; Keller, J. N., Abeta solubility and deposition during AD progression and in APPxPS-1 knock-in mice. *Neurobiol. Dis.* 2007, *27* (3), 301-11.

[19] Sultana, R.; Reed, T.; Perluigi, M.; Coccia, R.; Pierce, W. M.; Butterfield, D. A., Proteomic identification of nitrated brain proteins in amnestic mild cognitive impairment: a regional study. *J. Cell Mol. Med.* 2007, *11* (4), 839-51.

[20] Williams, T. I.; Lynn, B. C.; Markesbery, W. R.; Lovell, M. A., Increased levels of 4-hydroxynonenal and acrolein, neurotoxic markers of lipid peroxidation, in the brain in Mild Cognitive Impairment and early Alzheimer's disease. *Neurobiol. Aging* 2006, *27* (8), 1094-9.

[21] Butterfield, D. A.; Reed, T. T.; Perluigi, M.; De Marco, C.; Coccia, R.; Keller, J. N.; Markesbery, W. R.; Sultana, R., Elevated levels of 3-

nitrotyrosine in brain from subjects with amnestic mild cognitive impairment: implications for the role of nitration in the progression of Alzheimer's disease. *Brain Res.* 2007, *1148*, 243-8.

[22] Guidi, I.; Galimberti, D.; Lonati, S.; Novembrino, C.; Bamonti, F.; Tiriticco, M.; Fenoglio, C.; Venturelli, E.; Baron, P.; Bresolin, N.; Scarpini, E., Oxidative imbalance in patients with mild cognitive impairment and Alzheimer's disease. *Neurobiol. Aging* 2006, *27* (2), 262-9.

[23] Rinaldi, P.; Polidori, M. C.; Metastasio, A.; Mariani, E.; Mattioli, P.; Cherubini, A.; Catani, M.; Cecchetti, R.; Senin, U.; Mecocci, P., Plasma antioxidants are similarly depleted in mild cognitive impairment and in Alzheimer's disease. *Neurobiol. Aging* 2003, *24* (7), 915-9.

[24] Sultana, R.; Piroddi, M.; Galli, F.; Butterfield, D. A., Protein levels and activity of some antioxidant enzymes in hippocampus of subjects with amnestic mild cognitive impairment. *Neurochem. Res.* 2008, *33* (12), 2540-6.

[25] Migliore, L.; Fontana, I.; Trippi, F.; Colognato, R.; Coppede, F.; Tognoni, G.; Nucciarone, B.; Siciliano, G., Oxidative DNA damage in peripheral leukocytes of mild cognitive impairment and AD patients. *Neurobiol. Aging* 2005, *26* (5), 567-73.

[26] Wang, J.; Markesbery, W. R.; Lovell, M. A., Increased oxidative damage in nuclear and mitochondrial DNA in mild cognitive impairment. *J. Neurochem.* 2006, *96* (3), 825-32.

[27] Markesbery, W. R.; Kryscio, R. J.; Lovell, M. A.; Morrow, J. D., Lipid peroxidation is an early event in the brain in amnestic mild cognitive impairment. *Ann. Neurol.* 2005, *58* (5), 730-5.

[28] Ho, L.; Sharma, N.; Blackman, L.; Festa, E.; Reddy, G.; Pasinetti, G. M., From proteomics to biomarker discovery in Alzheimer's disease. *Brain Res. Brain Res. Rev.* 2005, *48* (2), 360-9.

[29] Jung, S. M.; Lee, K.; Lee, J. W.; Namkoong, H.; Kim, H. K.; Kim, S.; Na, H. R.; Ha, S. A.; Kim, J. R.; Ko, J.; Kim, J. W., Both plasma retinol-binding protein and haptoglobin precursor allele 1 in CSF: candidate biomarkers for the progression of normal to mild cognitive impairment to Alzheimer's disease. *Neurosci. Lett.* 2008, *436* (2), 153-7.

[30] Lee, J. W.; Namkoong, H.; Kim, H. K.; Kim, S.; Hwang, D. W.; Na, H. R.; Ha, S. A.; Kim, J. R.; Kim, J. W., Fibrinogen gamma-A chain precursor in CSF: a candidate biomarker for Alzheimer's disease. *BMC Neurol.* 2007, *7*, 14.

[31] Owen, J. B.; Di Domenico, F.; Sultana, R.; Perluigi, M.; Cini, C.; Pierce, W. M.; Butterfield, D. A., Proteomics-determined differences in the concanavalin-A-fractionated proteome of hippocampus and inferior parietal lobule in subjects with Alzheimer's disease and mild cognitive impairment: implications for progression of AD. *J. Proteome Res.* 2009, *8* (2), 471-82.

[32] Reed, T.; Perluigi, M.; Sultana, R.; Pierce, W. M.; Klein, J. B.; Turner, D. M.; Coccia, R.; Markesbery, W. R.; Butterfield, D. A., Redox proteomic identification of 4-hydroxy-2-nonenal-modified brain proteins in amnestic mild cognitive impairment: insight into the role of lipid peroxidation in the progression and pathogenesis of Alzheimer's disease. *Neurobiol. Dis.* 2008, *30* (1), 107-20.

[33] Simonsen, A. H.; McGuire, J.; Hansson, O.; Zetterberg, H.; Podust, V. N.; Davies, H. A.; Waldemar, G.; Minthon, L.; Blennow, K., Novel panel of cerebrospinal fluid biomarkers for the prediction of progression to Alzheimer dementia in patients with mild cognitive impairment. *Arch Neurol.* 2007, *64* (3), 366-70.

[34] Perkins, D. N.; Pappin, D. J.; Creasy, D. M.; Cottrell, J. S., Probability-based protein identification by searching sequence databases using mass spectrometry data. *Electrophoresis* 1999, *20* (18), 3551-67.

[35] Edelman, G. M.; Wang, J. L., Binding and functional properties of concanavalin A and its derivatives. III. Interactions with indoleacetic acid and other hydrophobic ligands. *J. Biol. Chem.* 1978, *253* (9), 3016-22.

[36] Sultana, R.; Newman, S. F.; Huang, Q.; Butterfield, D. A., Detection of carbonylated proteins in two-dimensional sodium dodecyl sulfate polyacrylamide gel electrophoresis separations. *Methods Mol. Biol.* 2008, *476*, 153-63.

[37] Mosesson, M. W., Fibrinogen gamma chain functions. *J. Thromb Haemost* 2003, *1* (2), 231-8.

[38] Wassell, J., Haptoglobin: function and polymorphism. *Clin. Lab.* 2000, *46* (11-12), 547-52.

[39] Akiyama, H.; Barger, S.; Barnum, S.; Bradt, B.; Bauer, J.; Cole, G. M.; Cooper, N. R.; Eikelenboom, P.; Emmerling, M.; Fiebich, B. L.; Finch, C. E.; Frautschy, S.; Griffin, W. S.; Hampel, H.; Hull, M.; Landreth, G.; Lue, L.; Mrak, R.; Mackenzie, I. R.; McGeer, P. L.; O'Banion, M. K.; Pachter, J.; Pasinetti, G.; Plata-Salaman, C.; Rogers, J.; Rydel, R.; Shen, Y.; Streit, W.; Strohmeyer, R.; Tooyoma, I.; Van Muiswinkel, F. L.; Veerhuis, R.; Walker, D.; Webster, S.; Wegrzyniak, B.; Wenk, G.;

Wyss-Coray, T., Inflammation and Alzheimer's disease. *Neurobiol. Aging* 2000, *21* (3), 383-421.

[40] Koh, A.; da Silva, A. P.; Bansal, A. K.; Bansal, M.; Sun, C.; Lee, H.; Glogauer, M.; Sodek, J.; Zohar, R., Role of osteopontin in neutrophil function. *Immunology* 2007, *122* (4), 466-75.

[41] Wang, K. X.; Denhardt, D. T., Osteopontin: role in immune regulation and stress responses. *Cytokine Growth Factor Rev.* 2008, *19* (5-6), 333-45.

[42] Denhardt, D. T.; Noda, M.; O'Regan, A. W.; Pavlin, D.; Berman, J. S., Osteopontin as a means to cope with environmental insults: regulation of inflammation, tissue remodeling, and cell survival. *J. Clin. Invest* 2001, *107* (9), 1055-61.

[43] Standal, T.; Borset, M.; Sundan, A., Role of osteopontin in adhesion, migration, cell survival and bone remodeling. *Exp. Oncol.* 2004, *26* (3), 179-84.

[44] Hong, D. P.; Gozu, M.; Hasegawa, K.; Naiki, H.; Goto, Y., Conformation of beta 2-microglobulin amyloid fibrils analyzed by reduction of the disulfide bond. *J. Biol. Chem.* 2002, *277* (24), 21554-60.

[45] Eakin, C. M.; Miranker, A. D., From chance to frequent encounters: origins of beta2-microglobulin fibrillogenesis. *Biochim. Biophys. Acta* 2005, *1753* (1), 92-9.

[46] Hershko, A.; Leshinsky, E.; Ganoth, D.; Heller, H., ATP-dependent degradation of ubiquitin-protein conjugates. *Proc. Natl. Acad. Sci. USA* 1984, *81* (6), 1619-23.

[47] Perry, G.; Friedman, R.; Shaw, G.; Chau, V., Ubiquitin is detected in neurofibrillary tangles and senile plaque neurites of Alzheimer disease brains. *Proc. Natl. Acad. Sci. USA* 1987, *84* (9), 3033-6.

[48] Hoyer, S., Oxidative energy metabolism in Alzheimer brain. Studies in early-onset and late-onset cases. *Mol. Chem. Neuropathol.* 1992, *16* (3), 207-24.

[49] Messier, C.; Gagnon, M., Glucose regulation and brain aging. *J. Nutr. Health Aging* 2000, *4* (4), 208-13.

[50] Watson, G. S.; Craft, S., Modulation of memory by insulin and glucose: neuropsychological observations in Alzheimer's disease. *Eur. J. Pharmacol.* 2004, *490* (1-3), 97-113.

[51] Sultana, R.; Perluigi, M.; Butterfield, D. A., Oxidatively modified proteins in Alzheimer's disease (AD), mild cognitive impairment and animal models of AD: role of Abeta in pathogenesis. *Acta Neuropathol.* 2009, *in press*.

[52] Hamajima, N.; Matsuda, K.; Sakata, S.; Tamaki, N.; Sasaki, M.; Nonaka, M., A novel gene family defined by human dihydropyrimidinase and three related proteins with differential tissue distribution. *Gene* 1996, *180* (1-2), 157-63.

[53] Kato, Y.; Hamajima, N.; Inagaki, H.; Okamura, N.; Koji, T.; Sasaki, M.; Nonaka, M., Post-meiotic expression of the mouse dihydropyrimidinase-related protein 3 (DRP-3) gene during spermiogenesis. *Mol. Reprod. Dev.* 1998, *51* (1), 105-11.

[54] Castegna, A.; Aksenov, M.; Thongboonkerd, V.; Klein, J. B.; Pierce, W. M.; Booze, R.; Markesbery, W. R.; Butterfield, D. A., Proteomic identification of oxidatively modified proteins in Alzheimer's disease brain. Part II: dihydropyrimidinase-related protein 2, alpha-enolase and heat shock cognate 71. *J. Neurochem.* 2002, *82* (6), 1524-32.

[55] Sultana, R.; Boyd-Kimball, D.; Poon, H. F.; Cai, J.; Pierce, W. M.; Klein, J. B.; Merchant, M.; Markesbery, W. R.; Butterfield, D. A., Redox proteomics identification of oxidized proteins in Alzheimer's disease hippocampus and cerebellum: an approach to understand pathological and biochemical alterations in AD. *Neurobiol. Aging* 2006, *27* (11), 1564-76.

[56] Adams, J. C., Formation of stable microspikes containing actin and the 55 kDa actin bundling protein, fascin, is a consequence of cell adhesion to thrombospondin-1: implications for the anti-adhesive activities of thrombospondin-1. *J. Cell Sci.* 1995, *108 (Pt 5)*, 1977-90.

[57] Adams, J. C., Roles of fascin in cell adhesion and motility. *Curr. Opin. Cell Biol.* 2004, *16* (5), 590-6.

[58] Yamashiro, S.; Yamakita, Y.; Ono, S.; Matsumura, F., Fascin, an actin-bundling protein, induces membrane protrusions and increases cell motility of epithelial cells. *Mol. Biol. Cell* 1998, *9* (5), 993-1006.

[59] Pinkus, G. S.; Lones, M. A.; Matsumura, F.; Yamashiro, S.; Said, J. W.; Pinkus, J. L., Langerhans cell histiocytosis immunohistochemical expression of fascin, a dendritic cell marker. *Am. J. Clin. Pathol.* 2002, *118* (3), 335-43.

[60] Butterfield, D. A.; Hensley, K.; Cole, P.; Subramaniam, R.; Aksenov, M.; Aksenova, M.; Bummer, P. M.; Haley, B. E.; Carney, J. M., Oxidatively induced structural alteration of glutamine synthetase assessed by analysis of spin label incorporation kinetics: relevance to Alzheimer's disease. *J. Neurochem.* 1997, *68* (6), 2451-7.

[61] Castegna, A.; Aksenov, M.; Aksenova, M.; Thongboonkerd, V.; Klein, J. B.; Pierce, W. M.; Booze, R.; Markesbery, W. R.; Butterfield, D. A.,

Proteomic identification of oxidatively modified proteins in Alzheimer's disease brain. Part I: creatine kinase BB, glutamine synthase, and ubiquitin carboxy-terminal hydrolase L-1. *Free Radic. Biol. Med.* 2002, *33* (4), 562-71.

[62] Hensley, K.; Hall, N.; Subramaniam, R.; Cole, P.; Harris, M.; Aksenov, M.; Aksenova, M.; Gabbita, S. P.; Wu, J. F.; Carney, J. M.; et al., Brain regional correspondence between Alzheimer's disease histopathology and biomarkers of protein oxidation. *J. Neurochem.* 1995, *65* (5), 2146-56.

[63] Coyle, J. T.; Price, D. L.; DeLong, M. R., Alzheimer's disease: a disorder of cortical cholinergic innervation. *Science* 1983, *219* (4589), 1184-90.

[64] Davies, M. J.; Fu, S.; Wang, H.; Dean, R. T., Stable markers of oxidant damage to proteins and their application in the study of human disease. *Free Radic. Biol. Med.* 1999, *27* (11-12), 1151-63.

[65] Perry, E. K.; Curtis, M.; Dick, D. J.; Candy, J. M.; Atack, J. R.; Bloxham, C. A.; Blessed, G.; Fairbairn, A.; Tomlinson, B. E.; Perry, R. H., Cholinergic correlates of cognitive impairment in Parkinson's disease: comparisons with Alzheimer's disease. *J. Neurol. Neurosurg. Psychiatry* 1985, *48* (5), 413-21.

[66] Wevers, A.; Witter, B.; Moser, N.; Burghaus, L.; Banerjee, C.; Steinlein, O. K.; Schutz, U.; de Vos, R. A.; Steur, E. N.; Lindstrom, J.; Schroder, H., Classical Alzheimer features and cholinergic dysfunction: towards a unifying hypothesis? *Acta Neurol. Scand. Suppl.* 2000, *176*, 42-8.

[67] Bader Lange, M. L.; Cenini, G.; Piroddi, M.; Abdul, H. M.; Sultana, R.; Galli, F.; Memo, M.; Butterfield, D. A., Loss of phospholipid asymmetry and elevated brain apoptotic protein levels in subjects with amnestic mild cognitive impairment and Alzheimer disease. *Neurobiol. Dis.* 2008, *29* (3), 456-64.

[68] Cutler, R. G.; Kelly, J.; Storie, K.; Pedersen, W. A.; Tammara, A.; Hatanpaa, K.; Troncoso, J. C.; Mattson, M. P., Involvement of oxidative stress-induced abnormalities in ceramide and cholesterol metabolism in brain aging and Alzheimer's disease. *Proc. Natl. Acad. Sci. USA* 2004, *101* (7), 2070-5.

[69] Geula, C.; Nagykery, N.; Nicholas, A.; Wu, C. K., Cholinergic neuronal and axonal abnormalities are present early in aging and in Alzheimer disease. *J. Neuropathol. Exp. Neurol.* 2008, *67* (4), 309-18.

[70] Chowdhury, I.; Mo, Y.; Gao, L.; Kazi, A.; Fisher, A. B.; Feinstein, S. I., Oxidant stress stimulates expression of the human peroxiredoxin 6 gene

by a transcriptional mechanism involving an antioxidant response element. *Free Radic. Biol. Med.* 2009, *46* (2), 146-53.

[71] Ralat, L. A.; Manevich, Y.; Fisher, A. B.; Colman, R. F., Direct evidence for the formation of a complex between 1-cysteine peroxiredoxin and glutathione S-transferase pi with activity changes in both enzymes. *Biochemistry* 2006, *45* (2), 360-72.

[72] Renes, J.; de Vries, E. E.; Hooiveld, G. J.; Krikken, I.; Jansen, P. L.; Muller, M., Multidrug resistance protein MRP1 protects against the toxicity of the major lipid peroxidation product 4-hydroxynonenal. *Biochem. J.* 2000, *350 Pt 2*, 555-61.

[73] Sultana, R.; Butterfield, D. A., Oxidatively modified GST and MRP1 in Alzheimer's disease brain: implications for accumulation of reactive lipid peroxidation products. *Neurochem. Res.* 2004, *29* (12), 2215-20.

[74] Tchaikovskaya, T.; Fraifeld, V.; Urphanishvili, T.; Andorfer, J. H.; Davies, P.; Listowsky, I., Glutathione S-transferase hGSTM3 and ageing-associated neurodegeneration: relationship to Alzheimer's disease. *Mech. Ageing Dev.* 2005, *126* (2), 309-15.

[75] Perluigi, M. S., R.; Cenini, G.; Di Domenico, F.; Memo, M.; Pierce, W.M.; Coccia, R.; Butterfield, D.A., Redox proteomics identification of HNE-modified brain proteins in Alzheimers disease: Role of lipid peroxidation in AD pathogenesis. *Proteomics Clin. Appli.* 2009, *Accepted*.

[76] Calabrese, V.; Scapagnini, G.; Colombrita, C.; Ravagna, A.; Pennisi, G.; Giuffrida Stella, A. M.; Galli, F.; Butterfield, D. A., Redox regulation of heat shock protein expression in aging and neurodegenerative disorders associated with oxidative stress: a nutritional approach. *Amino Acids* 2003, *25* (3-4), 437-44.

[77] Doorn, J. A.; Maser, E.; Blum, A.; Claffey, D. J.; Petersen, D. R., Human carbonyl reductase catalyzes reduction of 4-oxonon-2-enal. *Biochemistry* 2004, *43* (41), 13106-14.

[78] Agarwal-Mawal, A.; Qureshi, H. Y.; Cafferty, P. W.; Yuan, Z.; Han, D.; Lin, R.; Paudel, H. K., 14-3-3 connects glycogen synthase kinase-3 beta to tau within a brain microtubule-associated tau phosphorylation complex. *J. Biol. Chem.* 2003, *278* (15), 12722-8.

[79] Dougherty, M. K.; Morrison, D. K., Unlocking the code of 14-3-3. *J. Cell Sci.* 2004, *117* (Pt 10), 1875-84.

[80] Takahashi, Y., The 14-3-3 proteins: gene, gene expression, and function. *Neurochem. Res.* 2003, *28* (8), 1265-73.

[81] Frautschy, S. A.; Baird, A.; Cole, G. M., Effects of injected Alzheimer beta-amyloid cores in rat brain. *Proc. Natl. Acad. Sci. USA* 1991, *88* (19), 8362-6.
[82] Layfield, R.; Fergusson, J.; Aitken, A.; Lowe, J.; Landon, M.; Mayer, R. J., Neurofibrillary tangles of Alzheimer's disease brains contain 14-3-3 proteins. *Neurosci. Lett.* 1996, *209* (1), 57-60.
[83] Burkhard, P. R.; Sanchez, J. C.; Landis, T.; Hochstrasser, D. F., CSF detection of the 14-3-3 protein in unselected patients with dementia. *Neurology* 2001, *56* (11), 1528-33.
[84] Butterfield, D. A.; Abdul, H. M.; Opii, W.; Newman, S. F.; Joshi, G.; Ansari, M. A.; Sultana, R., Pin1 in Alzheimer's disease. *J. Neurochem.* 2006, *98* (6), 1697-706.
[85] Gothel, S. F.; Marahiel, M. A., Peptidyl-prolyl cis-trans isomerases, a superfamily of ubiquitous folding catalysts. *Cell Mol. Life Sci.* 1999, *55* (3), 423-36.
[86] Lu, K. P.; Hanes, S. D.; Hunter, T., A human peptidyl-prolyl isomerase essential for regulation of mitosis. *Nature* 1996, *380* (6574), 544-7.
[87] Zhou, X. Z.; Kops, O.; Werner, A.; Lu, P. J.; Shen, M.; Stoller, G.; Kullertz, G.; Stark, M.; Fischer, G.; Lu, K. P., Pin1-dependent prolyl isomerization regulates dephosphorylation of Cdc25C and tau proteins. *Mol. Cell* 2000, *6* (4), 873-83.
[88] Pastorino, L.; Sun, A.; Lu, P. J.; Zhou, X. Z.; Balastik, M.; Finn, G.; Wulf, G.; Lim, J.; Li, S. H.; Li, X.; Xia, W.; Nicholson, L. K.; Lu, K. P., The prolyl isomerase Pin1 regulates amyloid precursor protein processing and amyloid-beta production. *Nature* 2006, *440* (7083), 528-34.
[89] Sultana, R.; Butterfield, D. A., Regional expression of key cell cycle proteins in brain from subjects with amnestic mild cognitive impairment. *Neurochem. Res.* 2007, *32* (4-5), 655-62.
[90] Thompson, J. E.; Hopkins, M. T.; Taylor, C.; Wang, T. W., Regulation of senescence by eukaryotic translation initiation factor 5A: implications for plant growth and development. *Trends Plant Sci.* 2004, *9* (4), 174-9.
[91] Condeelis, J., Elongation factor 1 alpha, translation and the cytoskeleton. *Trends Biochem. Sci.* 1995, *20* (5), 169-70.
[92] Tome, M. E.; Fiser, S. M.; Payne, C. M.; Gerner, E. W., Excess putrescine accumulation inhibits the formation of modified eukaryotic initiation factor 5A (eIF-5A) and induces apoptosis. *Biochem. J.* 1997, *328 (Pt 3)*, 847-54.

[93] Ling, M.; Merante, F.; Chen, H. S.; Duff, C.; Duncan, A. M.; Robinson, B. H., The human mitochondrial elongation factor tu (EF-Tu) gene: cDNA sequence, genomic localization, genomic structure, and identification of a pseudogene. *Gene* 1997, *197* (1-2), 325-36.

[94] Chang, R. C.; Wong, A. K.; Ng, H. K.; Hugon, J., Phosphorylation of eukaryotic initiation factor-2alpha (eIF2alpha) is associated with neuronal degeneration in Alzheimer's disease. *Neuroreport* 2002, *13* (18), 2429-32.

[95] Ding, Q.; Markesbery, W. R.; Chen, Q.; Li, F.; Keller, J. N., Ribosome dysfunction is an early event in Alzheimer's disease. *J. Neurosci.* 2005, *25* (40), 9171-5.

[96] Li, X.; An, W. L.; Alafuzoff, I.; Soininen, H.; Winblad, B.; Pei, J. J., Phosphorylated eukaryotic translation factor 4E is elevated in Alzheimer brain. *Neuroreport* 2004, *15* (14), 2237-40.

[97] Yang, Y.; Turner, R. S.; Gaut, J. R., The chaperone BiP/GRP78 binds to amyloid precursor protein and decreases Abeta40 and Abeta42 secretion. *J. Biol. Chem.* 1998, *273* (40), 25552-5.

[98] Falahatpisheh, H.; Nanez, A.; Montoya-Durango, D.; Qian, Y.; Tiffany-Castiglioni, E.; Ramos, K. S., Activation profiles of HSPA5 during the glomerular mesangial cell stress response to chemical injury. *Cell Stress Chaperones* 2007, *12* (3), 209-18.

[99] Hoozemans, J. J.; Veerhuis, R.; Van Haastert, E. S.; Rozemuller, J. M.; Baas, F.; Eikelenboom, P.; Scheper, W., The unfolded protein response is activated in Alzheimer's disease. *Acta Neuropathol.* 2005, *110* (2), 165-72.

[100] Katayama, T.; Imaizumi, K.; Sato, N.; Miyoshi, K.; Kudo, T.; Hitomi, J.; Morihara, T.; Yoneda, T.; Gomi, F.; Mori, Y.; Nakano, Y.; Takeda, J.; Tsuda, T.; Itoyama, Y.; Murayama, O.; Takashima, A.; St George-Hyslop, P.; Takeda, M.; Tohyama, M., Presenilin-1 mutations downregulate the signalling pathway of the unfolded-protein response. *Nat. Cell Biol.* 1999, *1* (8), 479-85.

[101] Jensen, P. H.; Hojrup, P.; Hager, H.; Nielsen, M. S.; Jacobsen, L.; Olesen, O. F.; Gliemann, J.; Jakes, R., Binding of Abeta to alpha- and beta-synucleins: identification of segments in alpha-synuclein/NAC precursor that bind Abeta and NAC. *Biochem. J.* 1997, *323 (Pt 2)*, 539-46.

[102] Boyd-Kimball, D.; Sultana, R.; Poon, H. F.; Lynn, B. C.; Casamenti, F.; Pepeu, G.; Klein, J. B.; Butterfield, D. A., Proteomic identification of proteins specifically oxidized by intracerebral injection of amyloid beta-

peptide (1-42) into rat brain: implications for Alzheimer's disease. *Neuroscience* 2005, *132* (2), 313-24.
[103] Beach, T. G.; Walker, R.; McGeer, E. G., Patterns of gliosis in Alzheimer's disease and aging cerebrum. *Glia* 1989, *2* (6), 420-36.
[104] Korolainen, M. A.; Auriola, S.; Nyman, T. A.; Alafuzoff, I.; Pirttila, T., Proteomic analysis of glial fibrillary acidic protein in Alzheimer's disease and aging brain. *Neurobiol. Dis.* 2005, *20* (3), 858-70.

In: Proteomics
Editor: Giselle C. Rancourt
ISBN: 978-1-61668-691-8

Chapter 8

MULTIDIMENSIONAL CHROMATOGRAPHY: AN ESSENTIAL TOOL FOR PROTEOMICS

Chiara Cavaliere, Eleonora Corradini, Patrizia Foglia, Piero Giansanti, Roberto Samperi and Aldo Laganà *

Department of Chemistry, SAPIENZA Università di Roma, Rome, Italy

ABSTRACT

The general strategy in proteomic research consists in sample preparation, protein or peptide separation, their identification, and data interpretation. A critical step is certainly protein or peptide separation. Since increasingly complicated biological structures are studied by mass spectrometry (MS), the need for more powerful and highly resolving separation methods is growing. Consequently, multidimensional separation techniques in combination with MS have emerged as a powerful tool for the large-scale proteomic analysis. Until recently, two dimensional gel electrophoresis (2-DE) was the technique most often used for protein separation. The limitations of 2-DE in detecting low abundance, very small or large proteins, basic and membrane/ hydrophobic ones, as well as difficulties with process automation, have forced researchers to look for other methods of protein separation, such as multidimensional liquid chromatography coupled to MS (MDLC-MS) or tandem MS (MDLC-MS/MS). MDLC combines two or more forms of

* Corresponding author: Dipartimento di Chimica, SAPIENZA Università di Roma, Box n° 34 - Roma 62, Piazzale Aldo Moro 5, 00185 Rome, Italy Phone: +39-06-49913679 Fax: +39-06-490631, e-mail: aldo.lagana@uniroma1.it

LC to increase the peak capacity, and thus the resolving power of separation, to better fractionate peptides prior to entering the mass spectrometer. In this chapter, we analyze status and recent developments of the MDLC experiments in their fundamental components. It describes a variety of separation modes that have been employed to achieve protein-level or peptide-level separation, including size exclusion chromatography, ion exchange chromatography, and reversed-phase chromatography. We also discuss the advantages and disadvantages of two different approaches that can be followed for the studies of proteomics: protein-level separation or peptide-level separation.

WHY MULTI DIMENTIONAL LIQUID CHROMATOGRAPHY IN PROTEOMIC ANALYSIS?

Proteins are molecular products of genes and play a central role in many biological processes. The protein expression is a function of cellular and environmental conditions and consequently it varies depending on time and under different conditions. Thus the proteins are directly responsible for all physiological and pathological processes and the study of these molecules is essential in the interest of a complete picture of a biological system and its relationship with the outside. Over the past two decades, mass spectrometry (MS) has become an important tool for the analysis of proteins [1,2]. One current method for the analysis of protein mixtures is proteolytic digestion followed by Liquid Chromatography-Tandem Mass Spectrometry (LC-MS/MS). This approach overcomes many difficulties associated with protein mixture MS analysis [3]. MS/MS analysis has been particularly effective because the data can be directly used to identify peptides and subsequently infer which proteins are in the mixture [4]. This type of approach for the analysis of protein mixtures is often referred to as "shotgun" proteomics. Because increasingly complicated biological structures are studied by MS/MS, the need for more powerful and highly resolving separation methods has grown. Infact, proteins are identified by mass-to-charge ratios of peptides and their fragments and sufficient separation is required for unambiguous identifications. Therefore proteins MS is closely linked to and depends largely on the separation technologies to simplify incredibly complex biological samples prior to analysis of the mass. Front-end separation is also required to detect low-abundance species that would otherwise be overshadowed by a higher abundance signal. Therefore, both accuracy and sensitivity of a mass spectrometric experiment rely on efficient separation. There is a very strong conceptual link between chemical separation and MS in which the latter is viewed as the mass resolution dimension of molecules separation [1]. Selection of appropriate separation methods is often the first step in designing the proteomic application. Two major approaches to separation widely used in proteomics are gel based and gel free. Two-dimensional polyacrylamide gel electrophoresis (2D PAGE) is the historic centerpiece of the gel-based separation methods [5-8]. There are many excellent reviews that cover 2D PAGE and gel-based approaches to proteomics [9-12]. Gel-based methods have been traditionally used with pulsed ionization MALDI instruments in which the protein band can be excised, digested, and off-line sampled with MALDI source [13]. The limitations of 2-DE in detecting low abundance proteins, very small or large proteins, as well as basic and membrane/hydrophobic proteins [14-16], as well as difficulties with automation of the process, have forced researchers to look for other methods of protein separation, such as multidimensional liquid chromatography (MDLC). MDLC is by no means a new concept and has a long history, it has been enjoying a renaissance in proteomics [5]. MDLC combines two or more forms of LC to increase the peak capacity, and thus the resolving power, of separations to better fractionate peptides prior to entering the mass spectrometer. Furthermore for adequate representation of the proteome, only multidimensional separation techniques can provide resolving capability of thousands of protein species and have proven to be superior to one-dimensional approaches. These techniques have emerged as a powerful tool for the large-scale analysis of such complex samples [17-20]. It better resolves peptides differing in charge and hydrophobicity to minimize ion suppression and improve ionization efficiency, and it simplifies the complexity of peptide ions entering into the mass spectrometer to minimize undersampling. This last aspect is important because the tandem MS process is driven by data-dependent-acquisition (DDA) and has a finite cycle time. A higher peak capacity and better resolving power improve the acquisition of data and can lead to a better representation of the proteins in the mixture and permit the identification of low-abundance proteins [17,18,21,22]. For example various techniques prefractionation orthogonal on the level of protein and peptide level have been utilized for the characterization of part of yeast proteome leading to the identification of thousands of proteins [23, 24].

PROTEINS OR PEPTIDES LEVEL SEPARATION?

The protein-level and peptide-level separations have relative advantages and disadvantages. Proteins are sensitive to precipitation upon exposure to high salt concentrations, to basic pH values, and organic solvents. Peptides, on the other hand, are relatively stable in solution and generally do not exhibit solubility issues. However, peptide-level separations also have limitations, including the scattering of tryptic peptides from a single parent protein into multiple fractions, which can potentially reduce protein identification scores. Furthermore for adequate representation of the proteome, only multidimensional separation techniques can provide resolving capability of thousands of protein species and have proven to be superior to one-dimensional approaches.

MULTIDIMENSIONAL CHROMATOGRAPHIC TECHNIQUES

Currently in the literature are given a variety of multidimensional combinations that lead to an increase in the resolving power of the technique [25,26]. These methods can use different chromatographic techniques and a different number of dimensions. There are important factors to be considered as the amount of time required for analysis, compatibility with MS buffer used for the chromatographic separation, and the effective integration of two dimensions. Usually the last stage of separation, which usually is the step directly interfaced to a mass spectrometer, is the RPLC, which can provide high resolution, desalting of samples, and the compatibility of the phases with the ESI source and MS detection. The basis of RP method is the hydrophobic interaction between peptides and stationary phase. The stationary phase is the most common C18 covalently linked with a basic material of silica, these phases are called RP, C18, silica or octadecyl (better known as ODS). Peptides are loaded onto an RP column in a solution with a low content of organic phase, which allows on-line desalting and concentration at the same time. During the chromatographic run is gradually increased the amount of organic modifier in the mobile phase so that the peptides may elute according to the strength of hydrophobic interactions with the stationary phase. The peptides separation by RP chromatography has been widely studied in recent decades and significant progress has been made in this technique [27]. Because of its separation efficiency, superior to other LC techniques, and its excellent compatibility with ESI, RP remains an important method of peptides or protein separation. An important consideration for the development of multidimensional separations is the orthogonality of coupled techniques. The resolution can be maximized by combining chromatographic methods based on different principles of separation. While the RP chromatography is mainly used as second dimension in proteomics applications, a variety of chromatographic techniques were used for the first dimension. The most commonly used techniques are exclusion chromatography, strong cation exchange (SCX) and strong anion exchange [26,28-32]. Some factors are important in the first dimension, should have a large carrying capacity, to be configurable with the second dimension, and should use a solvent compatible with the second mode. Many of the methods listed above reflect these criteria, while others are more suited for off-line. One of the most widely used combinations for MDLC is SCX and RP. SCX keeps peptides based on electrostatic interactions. Sulfonyl end groups of the resin coat the surface and create a strong negative charge that is largely resistant to pH changes. Peptides are loaded onto an SCX column with a low pH buffer (3-4), which permits, block the dissociation of peptide's carboxylic groups and promote interactions between the protonated basic amino acid residues and the sulfonate groups of SCX resins. Peptides are eluted by increasing the strength of the salt buffer, which disrupts the interaction between peptides and sulfonate groups. To break the stronger interaction between peptide and SCX resin, the greater salt concentration is needed. Also experimentally showed that the phase SCX has additional features, such as hydrophobicity [33]. To minimize the hydrophobicity role and to facilitate the peptides denaturation, 10-15% of organic modifier is often added to the elution buffer. Fractionation of electrostatic interaction provides a degree of orthogonality to RP separation and therefore is an excellent complement. Multidimensional techniques discussed above can be coupled in off-line mode or in on-line mode. The first is the simplest and provides the fractions collection after the first dimension and a further separation of these with the second dimension interfaced to mass spectrometer. On-line mode, instead, is refers to a system in which the transfer of the analyte between the first and second dimension is automated and does not involve any disruption of flow [34]. To do this, generally a switching valve is placed between the two dimensions. The main advantages of these mode are the ease of automation and a reduced risk of sample loss and contamination than off-line mode. Use of switching valves often involves use of a intermediate column such as a trap for on-line sample desalting and this makes the configuration relatively flexible compared to the integrated column.

However, the passage of the sample in the switching valves and then exposure to surfaces and to additional connections can lead to loss of analyte. MDLC in on-line mode can also be performed using biphasic column. This integrated system is a simple system where the first 10-15 cm of the column is packed with RP material, followed by ~ 3-5 cm of SCX material. A final portion of RP can be added to act as desalting phase or as a further separation phase [35,36]. Peptides are loaded manually onto the column [37]. The manual loading of samples directly onto the column minimizes any loss of analyte that could occur through the valve system. The end of the capillary column RP usually forms a conical tip end so that the column serves as the ESI emitter, so as to have a minimum postseparation dead volume [38]. Multidimensional separation is achieved by passing a series of buffers in the column [28]. The peptides related to the SCX phase are first eluted with a 2-5 minutes pulse of saline solution and second are separated by a RP gradient. A second pulse of saline solution is then used to move another population of peptides on the RP column, this process is repeated a number of times. Because the sample transfer between phases occurs within a single section of the column, the dead volume becomes negligible. Generally, the on-line approaches are ideal when the sample available is limited and the losses must be minimized.

CONCLUSION

All techniques examined in this short communication were suitable for a proteomics study. However, each has specific advantages and limitations depending on the available equipment, expertise and monetary resources. By combining the most advanced LC systems and mass spectrometers currently available, researchers have significantly improved overall sensitivity and dynamic range, but performance is still limited relative to the needs of biomedical applications. MS based on MDLC approaches, however, offer significant promise for biological discovery. Besides in combination with other fractionation methods such as subcellular fractionation, immunodepletion and enrichment, or even SDS-PAGE, a significant amount of a proteome can be uncovered by this method.

REFERENCES

[1] Yates III, J.R. Mass spectral analysis in proteomics. *Annual Review of Biophysics and Biomolecular Structure,* 2004 33, 297–316.

[2] Aebersold, R. & Mann, M. (2003). Mass spectrometry-based proteomics. *Nature,* 422; 198–207.

[3] Washburn, MP; Wolters, D; Yates III, JR. Large-scale analysis of the yeast proteome by multidimensional protein identification technology. *Natural Biotechnology,* 2001 19, 242–247.

[4] Eng, J; McCormack, A; Yates, J. An Approach to Correlate Tandem Mass Spectral Data of Peptides with Amino Acid Sequences in a Protein Database. *Journal of the American Society for Mass Spectrometr,* 1994 5, 976–989.

[5] Iborra, F. & Buhler, JM. (1976). Protein subunit mapping. A sensitive high resolution method. *Analytical Biochemistry*, 74; 503–11.

[6] Klose, J. Protein mapping by combined isoelectric focusing and electrophoresis of mouse tissues. A novel approach to testing for induced point mutations in mammals. *Humangenetik,* 1975, 26, 231–43.

[7] Scheele, GA. Two-dimensional gel analysis of soluble proteins. Charaterization of guinea pig exocrine pancreatic proteins. *Journal of Biological Chemistry,* 1975, 250, 5375–5385.

[8] Klose, J. & Kobalz, U. (1995). Two-dimensional electrophoresis of proteins: an updated protocol and implications for a functional analysis of the genome. *Electrophoresis,* 16; 1034–59.

[9] Herbert, BR; Harry, JL; Packer, NH; Gooley, AA; Pedersen, SK; Williams, KL. What place for polyacrylamide in proteomics?. *Trends Biotechnology,* 2001 19, S3–S9.

[10] Monteoliva, L. & Albar, JP. (2004). Differential proteomics: an overview of gel and nongel based approaches. *Brief Function Genomic Proteomic,* 3; 220–39.

[11] Braun, RJ; Kinkl, N; Beer, M; Ueffing, M. Two-dimensional electrophoresis of membrane proteins. *Analytical Bioanalytical Chemistry,* 2007 389, 1033–45.

[12] Cooper, JW; Wang, Y; Lee, CS. Recent advances in capillary separations for proteomics. *Electrophoresis,* 2004 25, 3913–3926.

[13] Eckerskorn, C; Strupat, K; Karas, M; Hillenkamp, F; Lottspeich, F. Mass spectrometric analysis of blotted proteins after gel electrophoretic separation by matrix-assisted laser desorption/ionization. *Electrophoresis,* 1992 13, 664–65.

[14] Gor, A; Obermaier, C; Boguth, G; Harder, A; Scheibe, B; Wildgruber, R; Weiss, W. The current state of two-dimensional electrophoresis with immobilized pH gradients. *Electrophoresis,* 2000 21, 1037-1053.

[15] Gorg, A. & Weiss, W. (1999). Analytical IPG-Dalt. *Methods Molecular Biology,* 112; 189-195.

[16] Gygi, SP; Corthals, GL; Zhang, Y; Rochon, Y; Aebersold, R. Evaluation of two-dimensional gel electrophoresis-based proteome analysis technology. *Proceeding of the National Academy Science U.S.A.,* 2000 17, 9390-9395.

[17] Krijgsveld, J; Gauci, S; Dormeyer, W; Heck, AJ. In-gel isoelectric focusing of peptides as a tool for improved protein identification. *Journal of Proteome Research* 2006 5, 1721-1730.

[18] Reinders, J; Zahedi, R.P; Pfanner, N; Meisinger, C; Sickmann, A. Towards the complete yeast mitochondrial proteome: multidimensional separation techniques for mitochondrial proteomics. *Journal Proteome Research,* 2006 5, 1543-1554.

[19] Essader, AS; Cargile, BJ; Bundy, JL.; Stephenson, JL. A comparison of immobilized pH gradient isoelectric focusing and strong-cation-exchange chromatography as a first dimension in shotgun proteomics. *Proteomics,* 2005 5, 24-34.

[20] Lemeer, S; Pinkse, MW; Mohammed, S; van Breukelen, B; den Hertog, J; Slijper, M; Heck, A. J. Online Automated in Vivo Zebrafish Phosphoproteomics: From Large-Scale Analysis Down to a Single Embryo. *Journal Proteome Research,* 2008 7, 1555-1564.

[21] [21] Yaneva, M. & Tempst, P. (2006). Isolation and Mass Spectrometry of Specific DN Proteins. *Methods Molecular Biology,* 338; 291-303.

[22] Cargile, B.J; Sevinsky, J.R; Essader, AS; Stephenson, JL; Bundy, JL. Immobilized pH gradient isoelectric focusing as a first-dimension separation in shotgun proteomics. *.Journal Biomolecular Technology,* 2005 16, 181-189.

[23] de Godoy, LM; Olsen, J.V; Cox, J; Nielsen, ML; Hubner, NC; Frohlich, F; Walther, TC; Mann, M. Comprehensive mass *Nature,* 2008 455, 1251-1254.

[24] Wilson-Grady, JT; Villen, J; Gygi, SP. Phosphoproteome analysis of fission yeast. *Journal of Proteome Research,* 2008 7, 1088-1097.

[25] Link, AJ; Eng, J; Schieltz, DM; Carmack, E; Mize, GJ; Morris, DR; Garvik, BM; Yates, JR. Direct analysis of protein complexes using mass spectrometry. *Nat. Biotechnol,* 1999 17, 676–682.

[26] Mawuenyega, KG; Kaji, H; Yamauchi, Y; Shinkawa, T; Saito, H; Taoka, M; Takahashi, N; Isobe, T. Large-scale identi-fication of caenorhabditis elegans proteins by multidimensional liquid chromatography-tandem mass spectrometry. *J. Proteome Res,* 2003 2, 23–35.
[27] Claessens, H.A. & van Straten, M.A. (2004). A review on the chemical and thermal stability of stationary phases for reversed-phase liquid chromatography. *J. Chromatogr. A,* 1060; 23–41.
[28] Wolters, DA; Washburn, MP; Yates III, JR. An automated multidimensional protein identification technology for shotgun proteomics. *Anal. Chem,* 2001 73, 5683–5690.
[29] Machtejevas, E; John, H, Wagner, K. Automated multi-dimensional liquid chromatography: sample preparation and identification of peptides from human blood filtrate. *J. Chromatogr., B: Anal. Technol. Biomed. Life Sci,* 2004 803, 121–130.
[30] Peng, J; Elias, JE; Thoreen, CC; Licklider, LJ; Gygi, SP. Evaluation of Multidimensional chromatography coupled with tandem Mass Spectrometry (LC/LC-MS/MS) for Large-Scale Protein Analysis: The Yeast Proteome. *J. Proteome Res,* 2003 2, 43–50.
[31] de Godoy, LM; Olsen, JV; de Souza, GA; Li, G; Mortensen, P; Mann, M. Status of complete proteome analysis by mass spectrometry: SILAC labeled yeast as a model system. *Genome Biol,* 2006 7, R50.
[32] Chen, J;. Lee, CS; Shen, Y; Smith, RD; Baehrecke, EH. Integration of capillary isoelectric focusing with capillary reversed-phase liquid chromatography for two-dimensional proteomics separation. *Electrophoresis,* 2002 23, 3143–3148.
[33] Burke, TL; Mant, CT; Black' JA; Hodges, RS. Strong cation-exchange high-performance liquid chromatography of peptides: Effect of non-specific hydrophobic interactions and linearization of peptide retention behaviour. *J. Chromatogr,* 1989 476, 377–389.
[34] Dixon, SP; Pitfield, ID; Perrett, D. Comprehensive multi-dimensional liquid chromatographic separation in biomedical and pharmaceutical analysis: a review. *Biomed. Chromatogr,* 2006 20, 508– 529.
[35] McDonald, WH; Ohi, R; Miyamoto, DT; Mitchison, TJ; Yates III, JR. Comparison of three directly coupled HPLC MS/MS strategies for identification of proteins from complex mixtures: single-dimension LC-MS/MS, 2-phase MudPIT, and 3-phase MudPIT. *Int. J. Mass Spectrom,* 2002 219, 245–251.

[36] Wei, J; Sun, J; Yu, W; Jones, A; Oeller, P; Keller, M; Woodnutt, G; Short, JM. Global proteome discovery using an online threedimensional LC-MS/MS. *J. Proteome Res,* 2005 4, 801–808.

[37] Delahunty, C. & Yates III, JR. (2005). Protein identification using 2D-LC-MS/MS. *Methods,* 35; 248–255.

[38] Gatlin, CL; Kleeman, GR; Hays, LG; Link, AJ; Yates III, JR. Protein Identification at the Low Femtomole Level from Silver-Stained Gels Using a New Fritless Electrospray Interface for Liquid Chromotography-Microspray and Nanospray Mass Spectrometry. *Anal. Biochem,* 1998 263, 93–101.

In: Proteomics
Editor: Giselle C. Rancourt
ISBN: 978-1-61668-691-8

Commentary

PROTEOMIC APPROACH IN ANALYSING CARDIAC RESPONSES ON LOW-DOSE IONISING RADIATION USING CELLULAR AND TISSUE MODELS

*Soile Tapio**
Helmholtz Zentrum Muenchen – German Research Centre for Environmental Health, Institute of Radiation Biology, Neuherberg

It is well established that high doses of ionising radiation, such as used in radiotherapy, increase risk of cardiovascular diseases (CVD). Observed effects include direct damage to the coronary arteries, marked diffuse fibrotic damage of the pericardium and myocardium, pericardial adhesions, stenosis of the valves and microvascular damage [1, 2]. In contrast, there are considerable uncertainties concerning health effects of low doses of ionising radiation on heart. The need to explore potential biological and physiological effects at low doses is being increasingly acknowledged as the plans for new nuclear power plants and novel medical applications using low-dose radiation are emerging.

The data concerning CVD risk after occupational and environmental exposures to low doses of ionising radiation are controversial. Radiation workers in the Chernobyl liquidator cohort show increased risk for ischemic

* Helmholtz Zentrum Muenchen – German Research Centre for Environmental Health, Institute of Radiation Biology, Ingolstaedter Landstrasse 1, 85764 Neuherberg, soile.tapio@helmholtz-muenchen.de

heart disease [3]. Among employees at British Nuclear Fuels as well as in Canadian nuclear worker cohort and other occupationally radiation-exposed groups there is evidence for an increasing trend concerning circulatory disease mortality with dose [4, 5]. In contrast, no statistically significant increase in circulatory disease mortality due to inhaled radon or external γ-irradiation and its progenies could be observed in German uranium miners. However, while the risk for ischemic heart disease showed no increase, the rate of acute myocardial infarction was enhanced with radon dose [6].

The most convincing data showing excess radiation-associated risk for CVD has been observed in the Life Span Study of the Japanese atomic bomb survivors. Importantly, even at doses as low as 0.5 Gy the mortality and morbidity due to hypertension and myocardial infarction were increased [7, 8].

The risk of CVD with low and moderate doses of ionising radiation has recently been reviewed by Little et al. [9].

The vascular endothelium, a continuous monolayer containing of thin, flat cells located on the interior surface of blood vessels, forms an interface between circulating blood and subendothelial matrix [10]. It plays an important role in the integration and modulation of many functions of the arterial wall [11, 12]. Vascular endothelial dysfunction developing during the human aging process seems related to an increased production of reactive oxygen species (ROS) [13]. In atherosclerosis, increased endothelial production of ROS leads to oxidation of low density lipoproteins (LDL), accumulation of lipid into foam cells, intimal growth and finally atherosclerotic plaque expansion and rupture [14, 15].

Whether the biological responses of the endothelium in the case of high and low doses of ionising radiation are similar is still largely unknown. However, in contrast to high-dose radiation, acute doses in the range 0.1–1 Gy result in down-regulation of the adhesion of leukocytes to the endothelium both *in vitro* and *in vivo* and thus may have an anti-inflammatory effect [9, 16]. Furthermore, it is reasonable to believe that not only the dose but also the dose rate has an effect on the biological outcome [17].

We have used both a human endothelial cell line EA.hy926 and a mouse model to study the immediate proteomic effects of *in vitro* irradiation and long-term functional effects of heart-focussed *in vivo* irradiation, respectively.

As shown in a previous study, EA.hy926 retained most of the characteristics of primary endothelial cells (HUVEC) in a comparative cDNA expression profiling even after addition of statins that are used to reduce the risk of cardiovascular disease [18]. It may thus be considered as a good model system for the cardiac endothelium.

Ea.hy926 was irradiated with 0.2 Gy Co-60 gamma rays with two different dose rates (20 mGy/min and 200 mGy/min) and the cells were harvested 4h and 24 h after the irradiation. The proteome changes in the sham-irradiated vs. irradiated cytosolic fractions were analysed using 2 DE-DIGE techniques. Out of more than fifty protein spots that showed significant alterations in their expression 22 proteins were identified. Among the pathways affected by the low-dose ionising radiation are Ran and Rho/Rock pathways, stress response and glycolysis.

Furthermore, we found across this classification a group of proteins belonging to small Ras-like GTPases, namely Ran, RhoA and Sar1a that share a significant sequence homology, the GDP/GTP binding pocket being especially conserved [19]. Many Ras-superfamily small GTPases are components of signalling pathways that link extracellular signals via transmembrane G-protein-coupled receptors to cytoplasmic or nuclear responses [20]. Interestingly, the previous data show that both RhoA and Ran expression is dependent on the production of reactive oxygen species [21, 22].

Functional and proteomic alterations in mitochondria isolated from irradiated and sham-irradiated murine (C57Bl6) hearts 4 weeks after heart-focussed irradiation (0.2 Gy, 2 Gy X-ray) were analysed. No differences between sham- and irradiated cardiac mitochondria were found in swelling, respiratory coupling and production of ATP. However, we could identify significantly increased ROS formation in cardiac mitochondria 4 weeks after the exposure to 2 Gy ionising radiation. The results with the irradiated murine hearts emphasize the importance of the persistant oxidative stress as a result of low and moderate doses of ionising radiation. These results are in concordance with a large number of recent data suggesting that altered levels of oxidative stress are essential in the development of cardiovascular disease and that cardiac mitochondria may play an important role both as a target and source of reactive oxygen species [23-27].

Taken together, these results demonstrate the importance of proteomics in finding new biological target molecules of a low-dose radiation response in a critical tissue.

REFERENCES

[1] Demirci, S., Nam, J., Hubbs, J. L., Nguyen, T., Marks, L. B., Radiation-induced cardiac toxicity after therapy for breast cancer: interaction

between treatment era and follow-up duration. *Int J Radiat Oncol Biol Phys* 2009, *73*, 980-987.

[2] Adams, M. J., Hardenbergh, P. H., Constine, L. S., Lipshultz, S. E., Radiation-associated cardiovascular disease. *Crit. Rev. Oncol. Hematol.* 2003, *45*, 55-75.

[3] [3] Ivanov, V. K., Maksioutov, M. A., Chekin, S. Y., Petrov, A. V., *et al.*, The risk of radiation-induced cerebrovascular disease in Chernobyl emergency workers. *Health Phys.* 2006, *90*, 199-207.

[4] McGeoghegan, D., Binks, K., Gillies, M., Jones, S., Whaley, S., The non-cancer mortality experience of male workers at British Nuclear Fuels plc, 1946-2005. *Int. J. Epidemiol.* 2008, *37*, 506-518.

[5] Ashmore, J. P., Krewski, D., Zielinski, J. M., Jiang, H., *et al.*, First analysis of mortality and occupational radiation exposure based on the National Dose Registry of Canada. *Am. J. Epidemiol.* 1998, *148*, 564-574.

[6] Kreuzer, M., Kreisheimer, M., Kandel, M., Schnelzer, M., *et al.*, Mortality from cardiovascular diseases in the German uranium miners cohort study, 1946-1998. *Radiat. Environ. Biophys.* 2006, *45*, 159-166.

[7] Preston, D. L., Shimizu, Y., Pierce, D. A., Suyama, A., Mabuchi, K., Studies of mortality of atomic bomb survivors. Report 13: Solid cancer and noncancer disease mortality: 1950-1997. *Radiat. Res.* 2003, *160*, 381-407.

[8] Yamada, M., Wong, F. L., Fujiwara, S., Akahoshi, M., Suzuki, G., Noncancer disease incidence in atomic bomb survivors, 1958-1998. *Radiat. Res.* 2004, *161*, 622-632.

[9] Little, M. P., Tawn, E. J., Tzoulaki, I., Wakeford, R., *et al.*, A systematic review of epidemiological associations between low and moderate doses of ionizing radiation and late cardiovascular effects, and their possible mechanisms. *Radiat. Res.* 2008, *169*, 99-109.

[10] Marsden, P. A., Goligorsky, M. S., Brenner, B. M., Endothelial cell biology in relation to current concepts of vessel wall structure and function. *J. Am. Soc. Nephrol.* 1991, *1*, 931-948.

[11] Luscher, T. F., Richard, V., Tschudi, M., Yang, Z. H., Boulanger, C., Endothelial control of vascular tone in large and small coronary arteries. *J Am Coll Cardiol* 1990, *15*, 519-527.

[12] Furchgott, R. F., Zawadzki, J. V., The obligatory role of endothelial cells in the relaxation of arterial smooth muscle by acetylcholine. *Nature* 1980, *288*, 373-376.

[13] Herrera, M. D., Mingorance, C., Rodriguez-Rodriguez, R., Sotomayor, M. A., Endothelial Dysfunction and Aging: an Update. *Ageing Res Rev* 2009.

[14] Ross, R., Atherosclerosis--an inflammatory disease. *N Engl J Med* 1999, *340*, 115-126.

[15] Falk, E., Fernandez-Ortiz, A., Role of thrombosis in atherosclerosis and its complications. *Am J Cardiol* 1995, *75*, 3B-11B.

[16] Rodel, F., Hantschel, M., Hildebrandt, G., Schultze-Mosgau, S., *et al.*, Dose-dependent biphasic induction and transcriptional activity of nuclear factor kappa B (NF-kappaB) in EA.hy.926 endothelial cells after low-dose X-irradiation. *Int. J. Radiat. Biol.* 2004, *80*, 115-123.

[17] Amundson, S. A., Lee, R. A., Koch-Paiz, C. A., Bittner, M. L., *et al.*, Differential responses of stress genes to low dose-rate gamma irradiation. *Mol Cancer Res* 2003, *1*, 445-452.

[18] Boerma, M., Burton, G. R., Wang, J., Fink, L. M., *et al.*, Comparative expression profiling in primary and immortalized endothelial cells: changes in gene expression in response to hydroxy methylglutaryl-coenzyme A reductase inhibition. *Blood Coagul Fibrinolysis* 2006, *17*, 173-180.

[19] Neuwald, A. F., Kannan, N., Poleksic, A., Hata, N., Liu, J. S., Ran's C-terminal, basic patch, and nucleotide exchange mechanisms in light of a canonical structure for Rab, Rho, Ras, and Ran GTPases. *Genome Res* 2003, *13*, 673-692.

[20] [Bhattacharya, M., Babwah, A. V., Ferguson, S. S., Small GTP-binding protein-coupled receptors. *Biochem. Soc. Trans.* 2004, *32*, 1040-1044.

[21] Heo, J., Redox regulation of Ran GTPase. *Biochem. Biophys. Res. Commun* 2008, *376*, 568-572.

[22] Heo, J., Campbell, S. L., Mechanism of redox-mediated guanine nucleotide exchange on redox-active Rho GTPases. *J. Biol. Chem.* 2005, *280*, 31003-31010.

[23] Landar, A., Zmijewski, J. W., Dickinson, D. A., Le Goffe, C., *et al.*, Interaction of electrophilic lipid oxidation products with mitochondria in endothelial cells and formation of reactive oxygen species. *Am J Physiol Heart Circ Physiol* 2006, *290*, H1777-1787.

[24] Zhang, D. X., Gutterman, D. D., Mitochondrial reactive oxygen species-mediated signaling in endothelial cells. *Am J Physiol Heart Circ Physiol* 2007, *292*, H2023-2031.

[25] Ballinger, S. W., Mitochondrial dysfunction in cardiovascular disease. *Free Radic. Biol Med.* 2005, *38*, 1278-1295.

[26] Zmijewski, J. W., Landar, A., Watanabe, N., Dickinson, D. A., *et al.*, Cell signalling by oxidized lipids and the role of reactive oxygen species in the endothelium. *Biochem. Soc. Trans.* 2005, *33*, 1385-1389.

[27] Davidson, S. M., Duchen, M. R., Endothelial mitochondria: contributing to vascular function and disease. *Circ. Res.* 2007, *100*, 1128-1141.

INDEX

2

4

A

B

C

D

E

F

G

H

I

J

K

L

M

N

O

P

Q

R

S